MANAGING WATER IN CENTRAL ASIA

The Central Asian and Caucasian Prospects project is sponsored by:

- A Meredith Jones & Co. Ltd
- BG plc
- British Aerospace
- BP Amoco plc
- ENI S.p.A.
- Exxon/Mobil
- Shell International Petroleum Company Ltd
- Statoil

Series editor: Edmund Herzig
Head, Russia and Eurasia Programme: Roy Allison

CENTRAL ASIAN AND CAUCASIAN PROSPECTS

MANAGING WATER IN CENTRAL ASIA

PHILIP MICKLIN

Published in Great Britain in 2000 by the Royal Institute of International Affairs,
Chatham House, 10 St James's Square, London SW1Y 4LE
(Charity Registration No. 208 223)

Distributed worldwide by The Brookings Institution, 1775 Massachusetts Avenue, NW,
Washington, DC 20036-2188, USA

ISBN 1 86203 000 6

Typeset in Times by Koinonia, Manchester
Printed and bound in Great Britain by Chameleon Press
Cover design by Youngs Design in Production

CONTENTS

ABOUT THE AUTHOR

Philip Micklin is a retired professor of geography from Western Michigan University, Kalamazoo, Michigan. His primary research interests are water management in the former USSR, with a special focus on the new states of Central Asia and the Aral Sea. Dr Micklin has spent considerable time in the former Soviet Union and Central Asia over the past 33 years. He has published numerous articles, a monograph and a coedited book on water management.

ACKNOWLEDGMENTS

I would like to thank Dr Edmund Herzig and others at the Royal Institute of International Affairs for their assistance in preparing this study. I also express my thanks to the participants in the Chatham House study group in January 2000. Their valuable and insightful comments on the initial draft were very helpful in preparing the final copy. Additionally, I want to acknowledge financial support for research upon which this study is based provided by the Lucia Harrison Endowment Fund of the Department of Geography, Western Michigan University and the Division of Scientific and Environmental Affairs of NATO.

May 2000 P.M.

ABBREVIATIONS AND ACRONYMS

ASBP	Aral Sea Basin Program (World Bank)
BVO	basin water management authority (part of ICWC)
EPT	Environmental Policy and Technology
GLAVGIDROMET	Main Administration of Hydrometeorology of Uzbekistan
ICAS	Interstate Council on the Problems of the Aral Sea Basin
ICCKU	Interstate Council for Kazakhstan, Kyrgyzstan and Uzbekistan
ICWC	Interstate Commission on Water Coordination
IFAS	International Fund for the Aral Sea
O&M	operation and maintenance
SDC	Sustainable Development Commission
TACIS	Technical Assistance to the Commonwealth of Independent States (European Union project)
UNDP	United Nations Development Programme
USAID	United States Agency for International Development
WARMAP	Water Resources Management and Agricultural Production in the Central Asian Republics (part of the TACIS project)
WUA	water user association

TRANSLITERATION SYSTEM

Where a standard English form of proper names or place names exists, that spelling has been retained. For transliteration of Russian to English, the system developed by the US Board on Geographic Names, as used in the journal *Soviet Geography: Review and Translation* (now *Post-Soviet Geography and Economics*), is employed.

SUMMARY

The Aral Sea basin of Central Asia extends over 1.8 million km^2 and has a population of nearly 50 million. Fresh water is the critical natural resource for this largely arid domain. The Amu Dar'ya and Syr Dar'ya rivers flow from the mountains across the lowland deserts to the Aral Sea, a huge lake at the centre of the basin. Most of the region's people live in oases along rivers surrounded by deserts, and irrigated agriculture is the dominant economic activity.

Irrigation has its roots in antiquity. It was primarily based on small-scale, family-run units that operated on traditions and customs developed over millennia. The Soviets destroyed this system, substituting large, state-owned farms. Local farmer-irrigators had essentially no control over the operation of these and little incentive to make them productive. Under the Soviet state, the use of irrigation expanded rapidly, as did water withdrawals for it, but agricultural productivity and water use efficiency declined.

By the end of the Soviet period, water management was in crisis. Irrigation had been expanded to the limit of water availability and so much water had been removed from the Amu Dar'ya and Syr Dar'ya that their flow into the Aral Sea fell to a fraction of its former level. The sea shrank rapidly, with a myriad of associated negative environmental, economic and social consequences. The new states of the Aral Sea basin face the daunting task of developing approaches, institutions and interstate cooperation to cope with the severe water management problems inherited from the USSR.

The aim of the present study is to provide an evaluation of water management in the Aral Sea basin. The focus is on the use of water for irrigation and the consequences of this. The problems of water use for industry, municipal or other purposes, as well as water quality issues, are beyond the scope of this paper.

1 INTRODUCTION

The Aral Sea basin lies in the heart of Central Asia (Map 1). It covers a vast area, estimated at nearly 1.8 million km^2.[1] The basin includes the territory of seven states: Uzbekistan, Turkmenistan, Kazakhstan, Afghanistan, Tajikistan, Kyrgyzstan and Iran. Within its boundaries are Kzyl-Orda and Chimkent *oblasts* in southern Kazakhstan; most of Kyrgyzstan, with the exception of the northern and northeastern territory (the drainage basins of Lake Issyk-Kul' and the Chu and Talas rivers); nearly all of Uzbekistan, with the exception of a part of the Ust-Urt Plateau in the far northeast of the country; all of Tajikistan; the northern part of Afghanistan; a small part of the extreme northeast of Iran; and all but the western fifth of Turkmenistan.

Lands that now constitute five of the seven basin states (Uzbekistan, Kazakhstan, Tajikistan, Turkmenistan and Kyrgyzstan) were part of the Russian empire and its successor, the Soviet Union, from the late 19th century until the collapse of the USSR in 1991. Eighty-three per cent of the basin was situated in the Soviet Union and over 90 per cent of river flow came from its territory. Afghanistan and Iran control the residual portion of the basin and together they contribute no more than 9 per cent of river discharge. Neither was ever part of the Soviet state nor the preceding Tsarist empire. In 1996 the population of the basin, not including the portion lying in Iran, was an estimated 45.2 million (37.3 million in the former Soviet Republics and 7.9 million in Afghanistan).[2]

Because of its geographical extent within the basin, economic might and political-military power, the Soviet Union dominated the water management agenda. Moscow made all critical water management decisions for the portion of the Aral Sea basin within Soviet borders. The Ministry of Reclamation and Water Management (Minvodkhoz) in

[1] Tsentr mezhdunarodnykh proyektov, Goskompriroda SSSR [Centre for International Projects, State Committee for the Environment, USSR], *Sovremennoye sostoyaniye prirody, naseleniya i khozyaystva basseyna Aral'skogo morya* [The current state of the nature, population and economy of the Aral Sea basin] (Moscow, 1991), p. 4.

[2] Tashkent Institute of Engineers of Irrigation and Agricultural Mechanization and the Aral Sea International Committee, 'The Mirzaev Report', May 1998, Table 1. This report, named after the principal author, was translated from an earlier Russian-language report published in September 1996 under the title 'Conceptual Strategy for Proper Development of the Water Ecology Relations in the Aral Sea Basin'. It grew out of a seminar of independent Central Asian experts devoted to finding new approaches to alleviating water management problems in the Aral Sea basin.

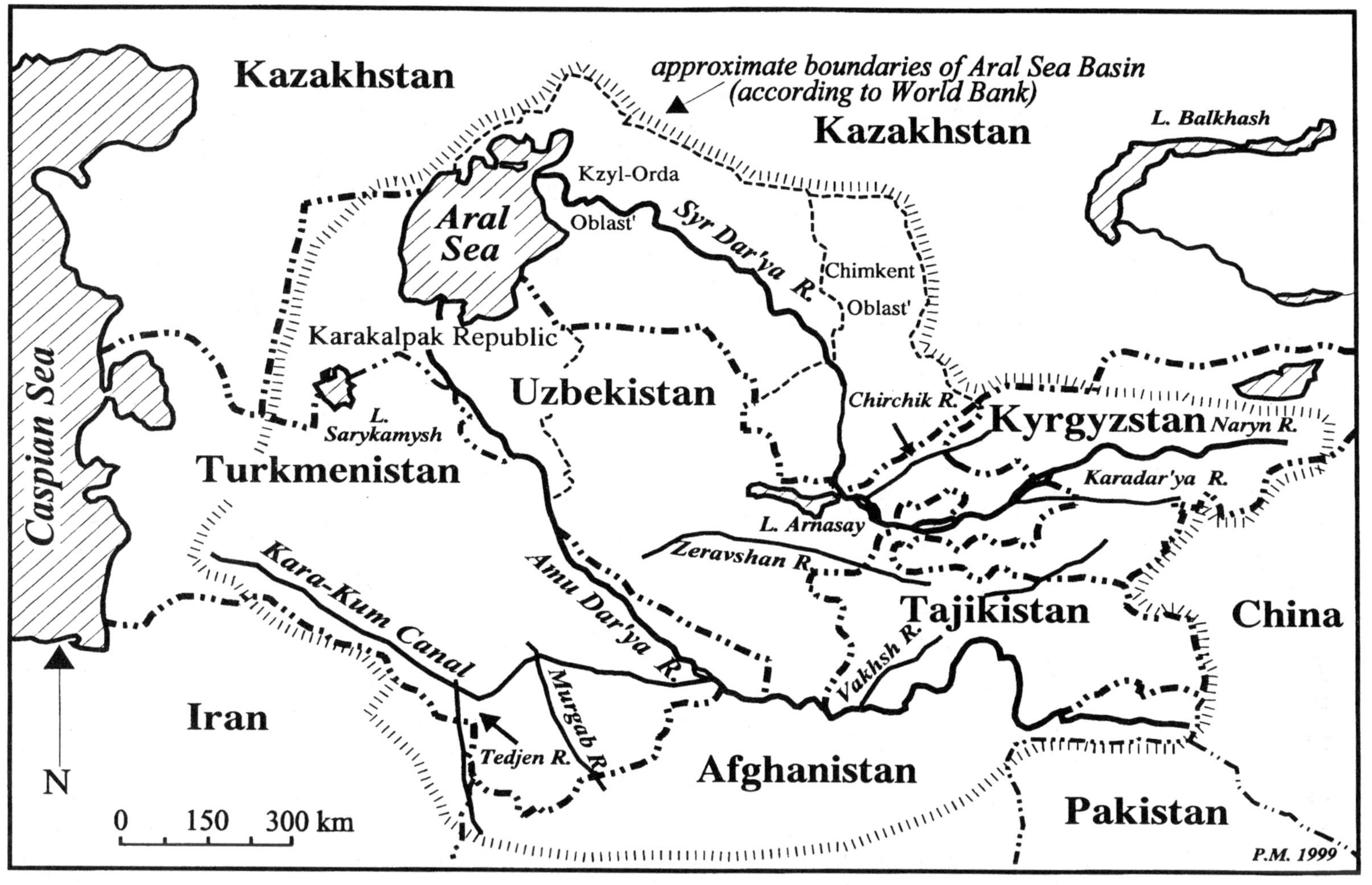

Map 1: The Aral Sea basin of Central Asia

Moscow settled disputes among the republics over the allocation and sharing of water from the international rivers, chiefly the Amu Dar'ya and Syr Dar'ya.[3] The Soviet Union paid little heed to Afghan and Iranian water management views or interests. Basin management was treated as essentially a domestic matter.

The situation changed dramatically at the end of 1991 with the shattering of the USSR. The Aral Sea basin and its water resources, rather than being controlled and managed by one superpower, were suddenly shared by seven states. None of these states is yet in a position, geographically, politically, economically or militarily, to dictate management policy for the basin on its own terms.

Nevertheless, geography and demography mean that certain states have much more interest in water management issues within the basin than others, as is evident from Table 1. All of Tajikistan and its population lie within the basin, as do 99 per cent of the territory and people of Uzbekistan. The basin accounts for close to 80 per cent of Turkmenistan and is home to nearly all the country's population. Over 70 per cent of Kyrgyzstan is in the basin and more than half its people reside here. In terms of geography and demography, these four states clearly have the pre-eminent stake in basin-wide water management issues. Kazakhstan has 13 per cent of its territory and 15 per cent of its population in the basin whereas Afghanistan has 40 per cent of its area in the basin with 33 per cent of its population there. These nations also have a significant interest in the basin. Only 2 per cent of Iranian territory, located in the extreme northeast of the country, is in the basin and it accounts for a minute portion of the national population (probably less than 1 per cent). Consequently, Iran has no significant interest in water management issues within the Aral Sea basin.

A somewhat different picture emerges when we look at the contribution the states make to the basin's area and population. Clearly dominant is Uzbekistan with 25 per cent of the area and 50 per cent of the population. Furthermore, Uzbekistan is in the middle of the basin (and Central Asia), is the only country with a border on five of the other six basin states (it has no border with Iran), and has a significant area and population in both of the river sub-basins of the Aral Sea drainage (the Syr Dar'ya in the north and east and the Amu Dar'ya in the south and west; see Map 1).

Turkmenistan and Kazakhstan both account for 21 per cent of the area, and 9 per cent and 5 per cent of the population, respectively; Afghanistan for 15 per cent of the basin area and 17 per cent of the population. Tajikistan and Kyrgyzstan both account for 8 per cent of the area but the former has 13 per cent of the population whereas the latter has only 5 per cent. Iran trails far behind the rest with 2 per cent of the basin area and probably no larger a share of the population. Obviously, other factors

[3] Yu. A. Kilinskiy and L.B. Sheynin, 'Pravovyye voprosy raspredeleniya resursov mezhrespublikansikh rek' [Legal questions on the distribution of resources of the rivers crossing inter-CIS borders], *Vodnyye resursy* [Water Resources], No. 3, 1986, p. 15.

Table 1: Geographic and demographic characteristics of the Aral Sea basin and riparian countries

State	Area (km^2)[a]	Area within Aral Sea Basin (km^2)[b]	% of total area of country within Aral Sea basin	% of Aral Sea basin area	Population (millions)[c]	Population in Aral Sea basin (millions)[d]	% of total population of country Aral Sea basin	% of Aral Sea basin population
Uzbekistan	447,232	438,287	98	25	23.784	23.664	99	50
Turkmenistan	488,680	378,000	77	21	4.298	4.255	99	9
Kazakhstan	2,728,185	365,400	13	21	16.847	2.546	15	5
Tajikistan	143,271	143,271	100	8	6.02	6.02	100	13
Kyrgyzstan	198,737	144,000	72	8	4.522	2.334	52	5
Afghanistan	653,004	262,800	40	15	24.792	8.128	33	17
Iran	1,640,015	34,200	2	2	68.96	See note e	NA	NA
Total all countries	6,299,123	1,765,958	28	100	149.223	46.948	31	100

[a] *World Almanac and Book of Facts, 1999* (Mahwah, NJ: Primedia, 1998), pp. 760–861.

[b] Measured by author from map in World Bank, *Aral Sea Basin Program (Kazakhstan, Kyrgyz Republic, Tajikistan, Turkmenistan and Uzbekistan)*: Water and Environmental Management Project, Washington, DC, May 1998.

[c] Mid-1998 estimates from *World Almanac and Book of Facts, 1999* (Mahwah, NJ: Primedia, 1998), pp. 760–861.

[d] Mid-1998 estimates derived from 1996 data in Tashkent Institute of Engineers of Irrigation and Agricultural Mechanization and the Aral Sea International Committee, 'The Mirzaev Report', September 1996, demographic appendices. Data for 1996 were increased by a (conservative) 1.5% average annual growth factor for two years to obtain the 1998 estimates.

[e] Information on the size of the Iranian population living within the Aral Sea basin is not available, but it is insignificant.

contribute to determining any country's influence within the basin, but geography and demography are key to understanding why Uzbekistan plays such a dominant role in water management within the basin.

The Aral Sea basin has great strategic importance. It is the heartland of Central Asia. One or more of the basin nations has borders with world powers China and Russia or with politically volatile Iran and Afghanistan. Three of the basin states are rich in fossil fuels – Kazakhstan with oil and Turkmenistan and Uzbekistan with natural gas. Uzbekistan is the world's third largest producer of cotton, nearly all of which is exported, and it has large reserves of gold and uranium. Since independence, these resources have garnered the attention and investments of the developed world.

Water is of critical importance in this largely arid region. Agriculture remains the dominant economic activity here and it is heavily dependent on extensive irrigation, requiring the application of huge quantities of water. Irrigation has strained the basin's water resources to the limit and led to disputes among the basin states over equitable water sharing. Irrigation has also resulted in the desiccation of the Aral Sea – once one of the world's most important lakes. The management of this key natural resource is of utmost importance to the economic, political and ecological future of the region.

In the following pages, I will analyse the complicated problems of managing water resources in the Aral Sea basin in the post-Soviet era. The primary focus is on the quantitative aspects of water withdrawn for irrigation and the consequences of this. The use of water for other purposes such as municipal and industrial supply, although an important issue, is beyond the scope of this study. Chapter 2 explores the region's water resource geography, the determination of which states possess the water resources and which states use them. The Aral Sea problem, which has attracted international attention to the region, is the subject of the Chapter 3. Chapter 4 looks at irrigated agriculture, which is at the heart of the most serious water management problems faced by the basin states. The critical issue of equitable sharing of the international water resources of the basin among the nations is discussed in Chapter 5. Chapter 6 delves into the internal (national) water policies and politics of the basin and looks at how these are affecting the process of decentralization and market reform in the management of water used for agricultural purposes. The final chapter summarizes the prospects for the future of water management in the Aral Sea basin.

2 BASIN WATER RESOURCES

The distribution of land and people in the Aral Sea basin is the key to understanding which nations are most concerned with water management issues in the region and the hierarchy of influence on these. Knowledge of the size, distribution and relative usage of water resources in the basin is also fundamental to elucidating and comprehending these problems.

The water resources of the Aral Sea basin may be divided into national and interstate (also known as transnational). The former consist of rivers, lakes, usable groundwater and return flows from uses situated entirely within the bounds of one or other of the basin states, that do not directly affect the other states. The latter are the same hydrologic entities that cross or form national borders or directly affect water resources in other basin nations.[1] In the Aral Sea basin, interstate water resources are far larger and more important than national water resources and will be the focus of attention here. The key transnational water resources of the basin are the two major rivers (Amu Dar'ya and Syr Dar'ya, with an aggregated drainage area of 1.7 million km^2) and the Aral Sea into which these rivers flow. The two rivers will be discussed in this chapter. The Aral Sea is such an important natural feature, has so many associated management problems and has been the recipient of such international attention that it deserves separate treatment (see Chapter 3).

2.1 Amu Dar'ya and Syr Dar'ya

The most important river within the Aral Sea basin is the Amu Dar'ya. Originating among the glaciers and snowfields of the Pamir mountains of Tajikistan, Kyrgyzstan and Afghanistan, it flows nearly 2,400 km from the mountains across the Kara-Kum desert and into the Aral Sea. During this journey, the river, or its major tributaries, flows along the borders of and across four states – Tajikistan, Afghanistan, Turkmenistan and Uzbekistan – entering, leaving, and re-entering the last two states several times (see Map 1). Average annual flow from the drainage basin is around 79 km^3. This includes not only the flow of the Amu Dar'ya and its tributaries but several

[1] World Bank and the ICWC, *Developing a Regional Water Management Strategy: Issues and Work Plan*, Aral Sea Basin Program Paper Series, Washington, DC, April 1996, p. 14.

'terminal' rivers[2] (Zeravshan, Murgab, Tedjen) that disappear in the deserts.[3] Of these rivers, all but the Kaskadar'ya cross international boundaries. The Amu Dar'ya is an 'exotic' river, which, in the hydrologic sense, means that all its flow originates in the well-watered Pamir mountains. This flow is substantially diminished by evaporation, transpiration from phreatophytic vegetation (deep-rooted plants that draw water from the zone of saturation) growing along its banks, and bed filtration as the river passes across the Kara-Kum desert to the Aral Sea. Owing to its exotic nature, even prior to the development of modern, large-scale irrigation, average inflow of the river to the Aral Sea decreased to around 40 km^3 from the 62 km^3 coming out of the mountains. Tajikistan contributes 80 per cent of flow generated in the Amu Dar'ya river basin, followed by Afghanistan (8 per cent), Uzbekistan (6 per cent), Kyrgyzstan (3 per cent) and Turkmenistan and Iran together around 3 per cent (most of which is formed in Iran).[4]

The Syr Dar'ya flows from the Tyan' Shan mountains, located to the north of the Pamirs. It too is chiefly fed by glaciers and snowmelt. With a total length of 2,500 km, it is somewhat longer than the Amu Dar'ya. The river (or its main tributaries the Naryn and Karadar'ya) flows from Kyrgyzstan into Uzbekistan, then across a narrow strip of Tajikistan that protrudes, thumb-like, into Uzbekistan and finally across Kazakhstan and into the Aral Sea. The average annual flow of the Syr Dar'ya, at 37 km^3, is considerably less than that of the Amu Dar'ya. Kyrgyzstan contributes 74 per cent of river flow, Uzbekistan 11 per cent, Kazakhstan 12 per cent and Tajikistan 3 per cent.[5] Like the Amu Dar'ya, the Syr Dar'ya is exotic. Prior to the modern age of irrigation, flow diminution was substantial during its long journey across the Kyzyl-Kum desert, with less than half (around 15 km^3 on an average annual basis) of the water coming from the mountains reaching the Aral Sea.

On an annual average basis, the two rivers (and the terminal rivers in the basin of the Amu Dar'ya) provide an estimated 116 km^3. Groundwater contributes additional water to river flow. According to recent estimates, total renewable groundwater

[2] Terminal rivers are not tributary to a body of water (river, lake, or sea). They are common in arid regions where they arise in humid mountainous zones and flow into deserts where evaporation rates are so high they lose all their water.

[3] Philip P. Micklin, *The Water Management Crisis in Soviet Central Asia*, The Carl Beck Papers in Russian and East European Studies, No. 905 (Pittsburgh PA: The Center for Russian and East European Studies, August 1991), p. 4; D. Mamatkanov, 'Vodnyye resursy gornoy territorii basseyna' [Water resources of the mountain terrritory basin], *Vestnik Arala* [Aral Bulletin], No. 1 (vodnyye resursy), spring 1996, pp. 5–7.

[4] D.C. McKinney and S. Akmansoy, 'What are the Competing Water Needs and Uses in the Aral Sea Region?', paper presented at Aral Sea Basin Workshop, sponsored by the SSRC, Tashkent Uzbekistan, May 19–21, 1998; ICAS, *Fundamental Provisions of Water Management in the Aral Sea Basin: A Common Strategy of Water Allocation, Rational Water Use and Protection of Water Resources*, prepared with the assistance of the World Bank, October 1996, Chapter 6.

[5] Ibid.

resources in the Aral Sea basin may be 44 km^3/yr, with perhaps 16 km^3/yr (36 per cent) usable.[6] Some 30 per cent of groundwater is believed to be transnational in nature, where the aquifer lies across national boundaries or is hydraulically connected to aquifers in other countries. Groundwater is a significant contributor to the flow of the Amu Dar'ya and the Syr Dar'ya in those rivers' headwaters. In the desert regions along the middle and lower courses, the rivers are net suppliers of flow to groundwater. As a result, it is difficult to ascertain the net addition made by groundwater to available water resources above and beyond the surface contribution to river flow.

During Soviet times, Central Asian water experts estimated usable groundwater that was not connected with river flow at 17 km^3/yr in the Aral Sea basin.[7] Using this figure, total basin water resources that are potentially usable (river flow plus groundwater that is not connected to river flow) amount on an average annual basis to about 133 km^3 (116 plus 17). The transnational part of groundwater resources would equal 5 km^3/yr if one applies the 30 per cent coefficient for the interstate portion cited above. The addition of this to the 116 km^3 for river flow provides an upper limit of 121 km^3 for transnational average annual water resources of the Aral Sea basin or 91 per cent of total usable basin water resources.

At 133 km^3/yr, the average annual water resources of the Aral Sea basin are substantial. On a per capita basis (assuming a mid-1998 basin population of nearly 47 million), they equal 2,830m^3/person. On a per unit area basis (assuming a basin area of 1.8 million km^2) they equal 74,000m^3/km^2. Per capita and per unit area figures are meaningless, however. They do not take into account the sharp spatial discontinuities of the region in terms of where flow is generated and where people live and use water most heavily. On this basis, we may divide the basin into two basic zones. The first comprises the upstream mountains where the flow is generated, which are sparsely inhabited and where water use is far lower than the available supply. This zone occupies only 20 per cent of the basin but generates 90 per cent of the flow for the Amu Dar'ya and Syr Dar'ya.[8] Second are the downstream arid plains (covering 80 per cent of the basin) where most of the population lives, where most of the water is used and where indigenous water resources are far smaller than usage. The deficit in the plains is, of course, covered by outflows from the well-watered mountains.

Tajikistan and Kyrgyzstan occupy the core of the mountain zone of the basin and are water-rich (see Figure 1). The former supplies 55 per cent of average annual basin river flow of 116 km^3 and the latter 25 per cent for an aggregate contribution of 80 per cent. Water withdrawals for the two countries together in 1995 were only 16 per cent

[6] ICAS, *Fundamental Provisions,* Chapter 6, Table 6.3. Usable groundwater is that portion of the total resource that has sufficiently low salinity and depth from the surface for it to be usable for drinking and economic purposes at a reasonable cost.

[7] Micklin, *The Water Management Crisis*, p. 99.

[8] Mamatkanov, 'Vodnyye resursy gornoy', p. 5.

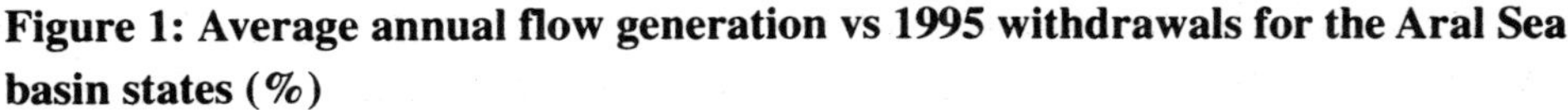

Figure 1: Average annual flow generation vs 1995 withdrawals for the Aral Sea basin states (%)

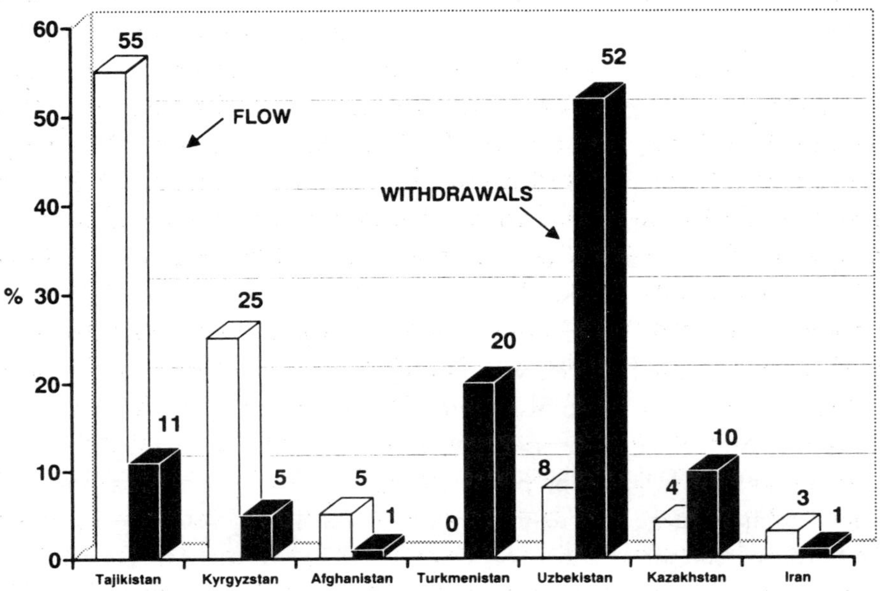

Sources: Compiled from World Bank, *Aral Sea Program (Kazakhstan, Kyrgyz Republic, Tajikistan, Turkmenistan and Uzbekistan)*, May 1998, Table 2; and other sources.

of the total. Consequently, these states are large net donors to basin water supplies. Afghanistan provides about 4 per cent of Aral Sea basin river flow but over 6 per cent for the Amu Dar'ya. Its withdrawals were much lower – probably not more than 1 per cent of the total in 1995, which also places it in the category of net upstream donor.[9]

The picture for the downstream states of Uzbekistan, Kazakhstan and Turkmenistan is the exact opposite. They are large net consumers of basin water resources. They lie mainly on the arid plains of the Central Asian deserts and contribute, as a group, only 14 per cent of Aral Sea basin river flow. Possessing substantial irrigated areas, these states took 83 per cent of estimated basin withdrawals (111 km^3) for all purposes in 1995. Uzbekistan contributes 8 per cent of basin flow but its withdrawals in 1995 were 52 per cent of the total. Turkmenistan contributes essentially no flow (most of the discharge of the Tedjen and Murgab rivers that enter Turkmen territory comes from Iran) but is a major consumer, accounting for 20 per cent of withdrawals in 1995. Kazakhstan contributes 4 per cent of aggregate basin flow, but 13 per cent of the flow for the Syr Dar'ya river, and withdrew 10 per cent of basin totals in 1995. Iran contributes about 3 per cent of basin flow and consumes, at most, 1 per cent.

[9] World Bank, et al., *Developing a Regional Water Management Strategy*, p. 16.

2.2 The sufficiency of renewable water resources

A key question for the management of water resources in the Aral Sea basin revolves around the sufficiency of the resource to meet demand. On an average annual basis, an upper limit estimate of renewable water resources in the basin (both national and transnational) is around 133 km^3 – 116 km^3 from the flow in the basins of the Amu Dar'ya and Syr Dar'ya, and 17 km^3 for groundwater not connected to river flow. For the most recent period of available data (1990–95), water withdrawals have ranged from a high of 126 km^3 in 1990 to a low of 111 km^3 in 1995.[10] These withdrawal figures are, at best, good estimates. There is some variation for them from one source to another, but they provide at least a reasonable idea of the range of withdrawals during the first half of the 1990s.

A portion of the flow withdrawn (estimated at near 24 km^3/yr for 1990–94) is returned to river channels, albeit with degraded quality, and is available for use downstream.[11] A large volume of withdrawn flow (estimated at 16 km^3/yr for 1990–94) is dumped into closed depressions in the deserts and evaporated. This is also potentially usable for irrigation purposes, although the salinity of some return flow is too high for this. Average total return flows are probably near 40 km^3 per year. One authoritative source estimates that 36–38 km^3 (90–95 per cent) of this is potentially available for reuse.[12] Including these reserves gives a total upper limit of near 170 km^3/yr for the usable water resources in the basin. This is significantly more water than is currently withdrawn and might suggest, at first glance, that there is plenty of water to go around for all basin countries and users, now and for the foreseeable future.

Unfortunately the situation is more complicated and less sanguine. First, there are losses of flow that are unavoidable, or at least difficult and costly to reduce (e.g. filtration from the river bed to surrounding land, evaporation from reservoirs, evapotranspiration from riparian vegetation).[13] Such losses may run to 16 km^3 in average flow years along the Amu Dar'ya and Syr Dar'ya.[14] Second, river flow is uneven on

[10] V. Dukhovnyy, *Melioratsiya i vodnoye khozyaystvo aridnogo zona* [Reclamation and water management of the arid zone] (Tashkent: Mekhnat, 1993), p 56; ICAS, *Fundamental Provisions*, Chapter 7, Tables 7.1 and 7.2.

[11] ICAS, *Fundamental Provisions*, Chapter 6, Table 6.6.

[12] ICAS, *Fundamental Provisions*, Chapter 6.

[13] This loss is primarily from water loving plants (phreatophytes) such as salt cedar, also known as tamarisk, (*gallica Linnaeus*), willow (*salix*), and cottonwood (*populus*) that grow along natural and artificial water courses in arid regions. These plants have deep roots and can draw huge amounts of water from significant depths to be lost through their leaves or other parts adapted to transpiring water to the atmosphere. Typical losses from these species in western America are 14,300–16,800 m^3/ha for salt cedar, 13,400 m^3/ha for willow, and 15,800–23,200 m^3/ha for cottonwood. Losses would be close to these in the climatically similar Aral Sea basin. Cotton, in desert regions of America, has water consumption rates from 7,000 to over 10,500 m^3/ha. See Frits Van der Leeden et al., *The Water Encyclopedia* (Chelsea, MI: Lewis Publishers, 2nd edn, 1990) pp. 117–24.

[14] ICAS, *Fundamental Provisions*, Chapter 7, Tables 7.1 and 7.2.

an intra- and inter-annual basis. There are seasons and years when flows are much higher than usage and others when they are much lower. To maximize the seasonal and multi-year availability of water, governments construct large dams and reservoirs to store water during high flow periods (spring and early summer) and high flow years, for use during summer low flow periods of high demand and low flow years.[15]

However, it is neither economically feasible nor environmentally wise to totally regulate rivers, particularly those as large as the Amu Dar'ya and Syr Dar'ya. Economically, the marginal cost of total or near total regulation (i.e. storing all or nearly all spring to early summer flow in every year for later release) would entail constructing costly additional storage capacity that would only be filled for short periods. Such an approach would also mean that for substantial periods during the high flow season river beds below the dams would be dry or nearly dry for significant distances (to the next major tributary or next reservoir) with extremely serious negative environmental and sanitary consequences.[16] Not all the seasonal surplus flow, and especially the surplus flow in high water years, can be stored for times when flow is low and demand is high.

Seasonal and multi-year storage dams have been built on both rivers and their tributaries to increase water resource availability during low flow periods. The aggregate, usable storage capacity in the entire Aral Sea basin is cited as 44 km^3 (17 in the basin of the Amu Dar'ya and 27 in the basin of the Syr Dar'ya).[17] The largest storage facilities are the Toktogul (gross capacity of 19.5 and usable capacity of 14 km^3) on the Syr Dar'ya, and the Nurek (gross capacity of 10.5 and usable capacity of 4.5 km^3) on the Vaksh, the main tributary of the Amu Dar'ya.[18] Storage has allowed the increase of the ensured yield of water (a measure of the flow that can be used) in a 90 per cent flow year,[19] occurring on average once in ten years, to 52 km^3 on the Amu Dar'ya and to 27 km^3 on the Syr Dar'ya – a total of 79 km^3. The amount of water that is available in the low flow years is, in fact, most crucial for water resource

[15] Micklin, *The Water Management Crisis*, pp. 4–7.

[16] M. Collier, R. Webb and J. Schmidt, *Dams and Rivers: Primer on the Ecological Effects of Dams*, US Geological Circular 1126 (Tucson, AZ: USGS, June 1996), 94 pages; Philip P. Micklin, 'Man and the water cycle: challenges for the 21st century,' *Geojournal*, Vol. 39, No. 3, July 1996, pp. 285–98.

[17] ICAS, *Fundamental Provisions*, Chapter 6. The full water storage volume of a reservoir is termed gross capacity while that portion of it that can be drained and refilled is known as usable capacity. The difference is termed dead storage.

[18] Dukhovnyy, *Melioratsiya i vodnoye*, p. 260; WARMAP project, *Formulation and Analysis of Regional Strategies on Land and Water Management*, July 1997, p. 8; A.N. Askochenskiy, *Orosheniye i obvodneniye v SSSR* [Irrigation and watering in the USSR] (Moscow: Kolos, 1967), p. 112; B.G. Shtepa (ed.), *Melioratsiya zemel' v SSSR* [Land Reclamation in the USSR] (Moscow: Kolos, 1975), p. 227.

[19] A 90 per cent flow year is a probabilistic concept. It is a flow year which probability analysis of a long record of annual flows, at least 30 years, indicates is likely to be exceeded 90 per cent of the time. The probability analyses used to create such probabilities are based on the fitting of a theoretical probability curve to the flow record or plotting of the actual flow record on probability paper.

management. It is more indicative of the state of water resources in arid regions such as the Aral Sea basin than the average annual figure, which is over-weighted by the high flow years, much of whose flow can be neither stored nor used.

When we examine water availability in low flow years, which usually occur in cycles in arid regions rather than being randomly distributed, the situation looks much less sanguine than the average annual flow scenario presented above. If we take the 79 km^3 figure for a 90 per cent flow year and subtract unavoidable losses of 16 km^3, only 64 km^3 remain as the usable resource. Assuming usable return flows of 38 km^3 and groundwater additions of 17 km^3 would give a total available resource of 119 km^3. This figure falls within the range of withdrawals (low of 111 km^3, high of 126 km^3 for the period 1990–95). But there are several caveats. First, to reach the 119 km^3 figure assumes two critical preconditions: the storage of nearly all spring high flows for later use, and the filling of the multi-year reservoirs to capacity at the beginning of the dry period. Second, full use would need to be made of usable groundwater and of return flows that do not reach rivers. In 1992, for example, estimates suggest that around 12 km^3 of the former (71 per cent) were used, and only 6 km^3 (38 per cent) of the latter.[20] On the basis of experience in the basin, these conditions are unlikely to be met at any time in the foreseeable future.

In reality, during low flow years withdrawals from the Amu Dar'ya and Syr Dar'ya river systems in the downstream net consuming countries of Uzbekistan, Kazakhstan, and Turkmenistan are, of necessity, substantially reduced (as are return flows). Furthermore, it is these years that cause heightened tensions among the states of the basin as the 'downstreamers' try to maximize their share of water coming from the 'upstreamers' (Tajikistan and Kyrgyzstan). The latter also strongly resist pressure from the former to increase the amount of water delivered downstream by reducing their usage and releasing more water from reservoirs on their territory.

As if water availability were not sufficiently problematic, the greatly diminished water supply to the Aral Sea engenders further severe difficulties, which are addressed in the next chapter.

[20] ICAS, *Fundamental Provisions*, Chapter 7, Table 7.2.

3 THE ARAL SEA AND ITS DESICCATION

3.1 Physical, hydrologic and biological character

The Aral Sea sits amidst the great deserts (Kara-Kum, Kyzyl-Kum, Betpakdala) of Central Asia (see Map 1). It is a terminal lake without surface outflow, and the balance between surface inflow from the Amu Dar'ya and Syr Dar'ya and net evaporation (evaporation from its surface minus precipitation on it) mainly determines the water level. Net groundwater exchange, which is difficult to measure, plays an insignificant role in the sea's water balance.[1] Inflow and net evaporation each averaged 56 km^3 from 1911 to 1960.[2] Hence, the water balance was in long-term equilibrium, with a maximum lake level variation over this period of less than one metre. Only two of the seven Aral Sea basin states (Kazakhstan and Uzbekistan) are riparian on the Aral Sea, each having an approximately equal length of shoreline. The entire Aral coastline within Uzbekistan lies within the Uzbek Karakalpakstan republic.

At nearly 67,000 km^2, the Aral Sea, according to area, was the world's fourth largest inland water body in 1960.[3] As a brackish lake with salinity averaging near 10 grammes/litre, less than one-third that of the ocean, it was inhabited chiefly by fresh-water species. The sea supported a major fishery and functioned as a key regional transportation route. The extensive deltas of the Syr Dar'ya and Amu Dar'ya sustained a diversity of flora and fauna. They also supported irrigated agriculture, animal husbandry, hunting and trapping, fishing and the harvesting of reeds. Over the past four decades the sea has steadily shrunk and salinized (see Figure 2 and Table 2). The main cause has been expanding irrigation that diminished inflow from the two influent rivers. The Aral divided into two water bodies in 1987 – a small Aral Sea in the north and a large Aral Sea in the south. The Syr Dar'ya flows into the former, and the Amu Dar'ya into the latter. A channel (river) has intermittently connected the two lakes. Between 1960 and 1998, the level of the small Aral fell by 13 metres and the

[1] V. N. Bortnik and S. P. Chistyaevaya (eds), *Gidrometeorologiya i gidrokhimiya morey SSSR* [Hydrometeorology and hydrochemistry of the seas of the USSR], Vol. VII, *Aral'skoye more* [Aral Sea] (Leningrad: Gidrometeoizdat, 1990), p. 38.

[2] Ibid., p. 36, Table 4.1; Philip P. Micklin, 'The Aral Sea Problem,' *Civil Engineering, Proceedings of the Institution of Civil Engineers*, August, 1994, p. 115.

[3] Micklin, *The Water Management Crisis*, pp. 42–54. The Caspian Sea in the Soviet Union and Iran, Lake Victoria in Africa and Lake Superior in America and Canada exceeded the Aral in surface area.

Figure 2: The shrinking Aral Sea

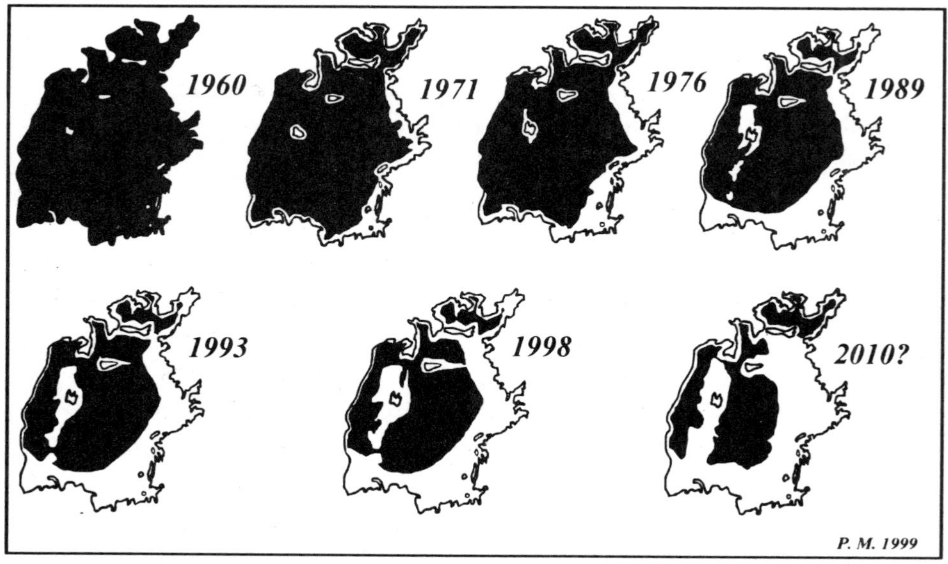

Sources: See Table 2.

large Aral by 18 metres. The area of both seas, taken together, diminished by more than 50 per cent and the volume by nearly 80 per cent. Salinity in the small sea rose by a factor of around three and in the large sea by a factor of nearly 4.5.

3.2 Ecological, economic and health consequences of desiccation

The anthropogenically caused desiccation of the Aral Sea has had a severe negative impact. However, this has not affected the basin states equally. Kazakhstan and Uzbekistan, as the shoreline riparians, have been most affected. Turkmenistan, although not abutting the sea, has some territory in the Amu Dar'ya delta, mainly Dashauz *oblast'*, that is close enough to have suffered substantial damage. Even within Kazakhstan, Uzbekistan and Turkmenistan, the territory suffering significant impact is a small part of each country and contains a minor portion of its population. For the three states combined, the affected area may cover, at most, 400,000 km^2, or 11 per cent of their aggregate territory. Some 3–4 million people out of a total population of 44 million (less than 10 per cent) live here.[4]

[4] Micklin, 'The Aral Sea Problem', p. 116.

Table 2: Hydrographic and hydrologic characteristics of the shrinking Aral Sea[a]

Year[b]	Level (m)	Area (km²)	% of 1960	Volume (km³)	% of 1960	Salinity (g/l)	% of 1960
1960	53.4	67,499	100	1089	100	9.9	100
Large Sea	53.4	61,381	100	1007	100		
Small Sea	53.4	6,118	100	82	100		
1971	51.1	60,200	89	925	85	11.6	113
1976	48.3	55,700	83	763	70	14.0	141
1989		39,734	59	364	33		
Large Sea	39.1	36,930	60	341	34	30.0	303
Small Sea	40.2	2,804	46	23	28	30.0	303
1993		35,424	52	297	27		
Large Sea	37.1	32,454	53	273	27	~35	353
Small Sea	40.8	2,970	49	24	29	~25	252
1998		31,541	47	236	22		
Large Sea	35.6	28,901	47	226	22	~45	454
Small Sea	39.5	2,640	43	21	25	~30	303
2010		22,144	33	148	14		
Large Sea	30.8	18,132	30	110	11	>60	606
Small Sea	45.0	4,012	66	38	46	~15	151

[a] Values from 1960 to 1998 are derived from Soviet data, data from Glavgidromet of Uzbekistan, data from S. Shivareva, Ye Ponenkova and B. Smerdov, 'Modelirovaniye urovnya Aral'skogo morya' [Modelling of the level of the Aral Sea], in *Problemy basseyna Aral'skogo Morya, issledovaniya, proyekty, predlozheniya* [Problems of the Aral Sea Basin, research, projects, proposals] (Tashkent: Chinor ENK, 1998), pp. 5–10 and calculations from a annualized water balance model developed by the author. Values for 2010 are calculated from the water balance model with the following assumptions: (1) average annual flow of Syr Dar'ya to Small Aral of 4.5 km³ for 1998–2010; (2) average annual flow of Amu Dar'ya to Large Aral of 16.5 km³ in 1998 and 5.5 km³ for 1999–2010; (3) flow from Small to Large Aral of 0 km³ for 1998–2005 and of 1.65 km³ for 2006–2010; (4) evaporation from the Small Aral of 0.960 m/yr and from the Large Aral of 0.966 m/yr; (5) precipitation on the surface of the Large Aral of 0.181m/yr and on the Small Aral of 0.198 m/yr.
[b] Values for 1960–76 are averages for the year; values for 1989–2010 are values on 1 January.

Furthermore, the administrative regions suffering the most (the Karakalpakstan republic in Uzbekistan, Kzyl-Orda *oblast'* in Kazakhstan and Dashauz *oblast'* in Turkmenistan) are politically impotent and do not have influence at the national level. Thus officials in Aslana, the new capital of Kazakhstan, and in Tashkent, capital of Uzbekistan, pay lip service to the view that the desiccation of the Aral and its associated consequences are a regional, and even global, catastrophe. (Such a view is less likely to be held in Ashgabat, capital of Turkmenistan, however.) This winning strategy has attracted substantial international aid. But these countries' own efforts and expenditures on solving the worst problems here have not matched their rhetoric precisely because the crisis zone is localized, has a relatively small population and lacks political clout.

The other states of the basin (Kyrgyzstan, Tajikistan, Afghanistan and Iran) are so distant from the zone where intense effects are apparent that they have suffered no demonstrable harm from the drying of the sea. These geographic considerations are key to understanding why Kazakhstan and Uzbekistan have been the motive forces in pushing the states of the basin to undertake measures to overcome or alleviate the worst problems, whereas Turkmenistan has shown much less interest and Kyrgyzstan and Tajikistan even less. Afghanistan and Iran have had no formal involvement whatsoever with Aral Sea issues.

The substantial Aral fishing industries developed by Kazakhstan and Uzbekistan in the first half of the 20th century ended in 1983 as the indigenous fish (20 species) which provided the basis for the commercial fishery disappeared as a result of rising salinity and the loss of shallow spawning and feeding areas.[5] All of the indigenous fish, however, survive in the deltaic lakes and Amu Dar'ya and Syr Dar'ya rivers – except the Aral salmon, which has become extinct. Four species of salt-tolerant fishes have been introduced and have survived; some have even flourished as competition from native species for food has disappeared. Even these, however, will probably vanish within years as a result of rising salinity, at least in the large Aral Sea. Several edible species remain, including Black Sea flounder, sprat and smelt, but are not caught commercially. Because of the loss of the fishery tens of thousands of people were thrown out of work. Navigation on the Aral also ceased as efforts to keep the increasingly long channels open to the major ports of Aral'sk at the northern end of the sea in Kazakhstan and Muynak at the southern end in Uzbekistan (Karakalpakstan) were abandoned.

[5] I. M. Zholdasova, L. P. Pavlovskaya, A. N. Urashbayev and S. K. Lubimova, 'Biological Bases of Fishery Development in the Waterbodies of the Southern Aral Region,' in *Ecological Research and Monitoring of the Aral Sea Deltas*, UNESCO Aral Sea Project 1992–6, Final Scientific Reports (Paris: UNESCO, 1998), p. 213–15; Micklin, *The Water Management Crisis*, pp. 49–50; N. V. Aladin and S. V. Kotov, 'Yestestvennoye sostoyaniye ekosistemy Aral'skogo morya i yeyo izmeneniye pri antropogennom vozdeystvii' [The natural condition of the Aral Sea ecosystem and changes to it owing to anthropogenic influence] in *Gidrobiologicheskiye problemy Aral'skogo morya* [Hydrobiologic problems of the Aral Sea], Trudy zoologicheskogo instituta, Vol. 199 (Leningrad), 1989, pp. 4–25; W.D. Williams and N.V. Aladin, 'The Aral Sea: recent limnological changes and their conservation significance', *Aquatic Conservation: Marine and Freshwater Ecosystems*, vol. 1 (1991), pp. 3–23. The last two works are excellent and detailed analyses of biological changes in the Aral Sea by leading experts on this subject.

The zone of significant damage stretches well beyond the sea and its immediate shoreline. Severe consequences are apparent in an approximately 400,000 km^2 region around the sea (including all of the Karakalpakstan republic and Khorezm *oblast'* in Uzbekistan, Dashauz *oblast'* in Turkmenistan, the western half of Kzyl-Orda and the southeastern portion of Akhtyubinsk *oblast'* in Kazakhstan) with a population of nearly 4 million. This region is commonly designated an 'ecological disaster zone' (*zona ekologicheskogo bedstviya*).

Damage has been particularly severe to the rich ecosystems of the extensive Amu Dar'ya delta, primarily located in the Karakalpak republic of Uzbekistan but stretching into Turkmenistan's Dashauz *oblast'*.[6] The Syr Dar'ya delta in Kzyl-Orda *oblast'* of Kazakhstan has also suffered. Greatly reduced river flows through the deltas, the virtual elimination of spring floods in the deltas (as a result of both reduced river flow and construction of upstream storage reservoirs) and declining groundwater levels (caused by the falling level of the Aral Sea) have led to spreading and intensifying desertification. Halophytes (plants tolerant of saline soils) and xerophytes (plants tolerant of dry conditions) are rapidly replacing endemic vegetation communities.[7] In some places, salts have accumulated on the surface forming *solonchak* (salt pans) where practically nothing will grow. Expanses of unique *tugay* (vegetation communities of trees, bushes and tall grasses, including poplar, willow, oleaster, salt cedar and reeds) that formerly stretched along all the main rivers and distributory channels here have been particularly hard hit. According to one expert, *tugay* covered 100,000 ha in the Amu Dar'ya delta in 1950, but shrank to 52,000 ha by the 1970s and to 15–20,000 ha by the mid-1990s.[8] *Tugay* complexes around the Aral Sea are habitats for a diversity of animals, including 60 species of mammals, more than 300 types of birds and 20 varieties of amphibians.[9]

Desiccation of the deltas has significantly diminished the area of lakes, wetlands and their associated reed communities. Between 1965 and 1986, the area of reeds in the Amu Dar'ya delta decreased from 500,000 ha to 1,000 ha.[10] This has resulted in

[6] Micklin, *The Water Management Crisis*, pp. 50–52.

[7] Micklin, 'The Aral Sea Problem', p. 116; for a detailed, scientific treatment of anthropogenic botanical changes in the Aral Sea region over the last few decades, see N. Novikova, *Printsipy sokhraneniya botanicheskogo raznoobraziya deltovykh ravnin Turana* [Principles of preserving the botanical diversity of the deltaic plains of Turan], dissertatsiya v forme nauchnogo doklada na soiskaniye uchenoy stepeni doktora geograficheskikh nauk [dissertation submitted for the degree of Doctor of Geographical Science] (Moscow, 1997), p. 104.

[8] N. Novikova, The Tugai of the Aral Sea Region is Dying: Can it be Restored?', *Russian Conservation News*, February 1996, No. 6, pp. 22–3.

[9] Ibid.

[10] M. Palvaniyazov, 'Vlyaniye pyl'nykh bur' na mestoobitaniya nekotorykh mlekopitayushchikh primorskoy zony Aral'skogo morya' [The influence of dust storms on the habitats of certain mammals of the coastal zone of the Aral Sea] *Problemy osvoyeniya pustyn'* [Problems of Desert Development], No. 1, 1989, p. 56.

serious ecological consequence as lakes, wetlands and their associated reed communities provide prime habitats for a variety of permanent and migratory waterfowl, a number of which are endangered.[11] Diminution of the aggregate water surface area coupled with increasing pollution of the remaining water bodies (primarily from irrigation run-off containing salts, fertilizers, pesticides, herbicides and cotton defoliants) has decimated aquatic bird populations.

Irrigated agriculture in the deltas of the Amu Dar'ya and Syr Dar'ya has suffered from an inadequacy of water as inflow to the deltas has decreased owing to heavy upstream consumptive use for irrigation. In addition, water that does reach the deltas has elevated salinity from the leaching of salts caused by repeated usage in the middle and upper courses of the rivers.[12] At times over 2 grammes/litre, these saline flows have lowered crop yields and, in conjunction with inadequate drainage of irrigated fields, promoted secondary soil salinization. Animal husbandry, both in the deltas and in desert regions adjacent to the Aral Sea, has been damaged by the reduction of the area and declining productivity of pastures resulting from desertification, dropping groundwater levels and replacement by inedible species of the natural vegetation that was suitable for grazing. For example, in the Amu Dar'ya delta, the area of natural pastures fell from 350,000 ha in the 1950s to 125,000 ha by the late 1980s, while the productivity of the remaining pastures was halved.[13]

Frequent strong winds are blowing sand, salt and dust onto adjacent lands from the dried bottom of the Aral Sea, now largely a barren, salt-covered desert with an area over 36,000 km^2. The major problem is the dust and salt that is transported great distances. Since the mid-1970s, satellite images have shown huge plumes extending 200 and even 400 km, allowing dust and salt to settle over a considerable area adjacent to the sea in Uzbekistan, Kazakhstan and, to a lesser degree, in Turkmenistan.[14] One scientist, however, contends that very light, fine particles of dust and salt are lifted several kilometres and travel 5–10,000 kilometres before settling back to earth.[15]

Some 60 per cent of the storms occur with north and northeast winds, which carry the dust and salt over the Ust-Urt plateau to the west of the sea and to the Amu Dar'ya delta at its southern end. The delta is the most densely settled area and economically

[11] Ibid.; Micklin, 'The Aral Sea Problem', p. 116;

[12] World Bank, *Aral Sea Basin Program*, pp. 3–5.

[13] Novikova, *Printsipy sokhraneniya botanicheskogo*, p. 71.

[14] Micklin, *The Water Management Crisis*, pp. 48–9; D.B Oreshkin, 'Aral'skaya katastrofa' [The Aral Catastrophe], *Nauka o zemle* [Land Science], No. 2 (Moscow: Znaniye, 1990); N.F. Glazovskiy, *Aral'skiy krizis: prichiny vozniknoveniya i put' vykhoda* [The Aral Crisis: Causes and Solutions] (Moscow: Nauka, 1990), pp. 20–23.

[15] A. A. Tursunov, 'Aral'skoye more i ekologicheskaya obstanovka v Sredney Azii i Kazakhstane' [The Aral Sea and the ecological situation in Central Asia and Kazakhstan], *Gidrotekhnicheskoye stroitel'stvo*, [Hydrotechnical construction], No. 6, 1989, p. 15.

and ecologically the important region around the sea.[16] Estimates of the total deflated material range from 13 million to as high as 231 million metric tonnes/yr, with the most probable value between 15 and 150 million tonnes.[17] The entrained salts are believed to constitute only 1 per cent of the total, so that the tonnage of salt transferred, even at the higher figure, would be 1.5 million tonnes.[18] However, a careful study, completed in the mid-1980s by well-known geologists and experts on the Aral, concluded that annual aeolian transport of salt alone from the dried bottom was around 43 million metric tonnes but would decrease slightly to 39 million metric tones by 2000.[19]

Whatever the true tonnage, considerable amounts of salts in dry and aerosol forms, the most harmful of which include sodium bicarbonate, sodium chloride and sodium sulphate, are settling on natural vegetation and crops.[20] In some cases, plants are killed outright but more commonly their growth (and for crops, yield) is substantially reduced. The salt and dust also have ill effects on wild and domestic animals, harming them directly and reducing their food supply.[21] Local health experts also consider airborne salt and dust to be a contributory factor in the high levels of respiratory illnesses and impairments, eye problems and possibly even throat and oesophageal cancer in the region.[22]

Owing to the shrinkage of the sea, climate has changed in a band up to 100 km wide along the former shoreline in Kazakhstan and Uzbekistan.[23] Maritime conditions have been replaced by more continental and desertic regimes. Summers have warmed and winters cooled, spring frosts are later and autumn frosts earlier, humidity is lower

[16] Bortnik and Chistyaevaya (eds), *Gidrometeorologiya i gidrokhimiya morey SSSR*, p. 27, Fig. 2.7.

[17] Glazovskiy, *Aral'skiy krizis*, p. 22.

[18] G. N. Chichasova (ed.), *Gidrometeorologicheskiye problemy Pryaral'ya* [Hydrometeorological problems of the near Aral region] (Leningrad: Gidrometeoizdat, 1990), p. 215.

[19] I. M. Rubanov and N. M. Bogdanava, 'Kolichestvennaya otsenka solevoy deflyatsii na osushayushchemsya dne Aral'skogo morya' [The quantitative aspect of saline deflation on the dried bottom of the Aral Sea] *Problemy osvoyeniya pustyn'*, No. 3, 1987, p. 14. The reason for the predicted decrease in exported dust and salt, in spite of a larger area of dried bottom, is an expected significant decrease in losses from 'old' areas of the bottom where the loose material subject to transport would become largely depleted, carried by precipitation below the surface or formed into a hard, deflation-resistant crust.

[20] M.Ye. Bel'gibayev, 'Pylesolemep – pribor dlya ulavlivaniya pyli i soley v vozdushnom potoke' [Dust/salt meter – Instrument for detecting dust and salt in an air stream], *Problemy osvoyeniya pustyn'*, No. 1, 1984, p. 74.

[21] Palvaniyazov, 'Vlyaniye pyl'nykh bur' na', p. 56.

[22] Ch. A. Abdirov, N. A. Agadzhanyan, A. V. Vervikhvost, K. P. Primbetov, A. Ye. Severin, Yu. P. Brushkov and L. G. Konstantinovaet, 'Stress reaktsiya zdorovykh detey g. Nukusa na vozdeystviya otritsatel'nykh ekologicheskikh faktorov v Pryaral'ye' [Stress reactions among healthy children in the town of Nukus to the influence of negative ecological factors in the near Aral region], *Vestnik Karakalpakskogo otdeleniya Akademii Nauk Respubliki Uzbekistana* [Bulletin of the Karakalpak section, Academy of Sciences of the Uzbekistan Republic], No. 2, 1993, pp. 15–20; *Pravda Vostoka*, 22 May 1987, p. 3; Tursunov, 'Aral'skoye more', pp. 15–16.

[23] Micklin, *The Water Management Crisis*, pp. 52–3; Glazovskiy, *Aral'skiy krizis*, pp. 19–21.

and the growing season shorter. The growing season has decreased by an average of ten days in the northern part of the Amu Dar'ya delta and become too short for cotton, forcing a switch from this crop to rice.

The population living in the ecological disaster zone suffers acute health problems.[24] Some of these, as indicated above, are direct consequences of the sea's recession (e.g. respiratory and digestive afflictions and possibly cancer, from inhalation and ingestion of blowing salt and dust and poorer diets from the loss of Aral fish as a major food source). Other serious health-related problems stem from environmental pollution associated with the heavy use of toxic chemicals in irrigated agriculture. However, probably the most serious health issues are directly related to Third World medical, health, nutrition and hygienic conditions and practices. Bacterial contamination of drinking water is pervasive and has led to very high rates of typhoid, paratyphoid, viral hepatitis and dysentery. Tuberculosis is prevalent, as is anaemia, particularly in pregnant woman. Liver and kidney ailments are widespread. Medical care is very poor, diets lack variety and adequate sewage systems are rare. General and infant mortality and morbidity are the highest among the states of the former Soviet Union.

Health conditions in the Karakalpak republic in Uzbekistan appear to be the worst in the Aral Sea basin. Data from surveys conducted in the mid- to late 1980s showed the average infant mortality rate to be more than 70 per 1,000 live births and in several districts it ranged from over 80 to over 100 per 1,000 live births. These rates are three to four times the national level in the former USSR and seven to ten times that of the United States.[25]

3.3 Restoration requirements

Although the severe impact of the drying of the Aral is felt over a relatively small portion of the basin (some 400,000 km^2 out of 1.8 million km^2, or 22 per cent), alleviation of the problem must involve the management of water over the entire Aral basin. Particularly important are the areas of greatest irrigation water usage in Turkmenistan, Kazakhstan and Uzbekistan. It is here that so much of the flow of the Amu Dar'ya and Syr Dar'ya is withdrawn for irrigation and consumptive use. This has led to a great reduction in discharge to the sea.[26]

[24] Philip P. Micklin, 'The Aral Crisis: Introduction to the Special Issue', *Post-Soviet Geography*, Vol. 33, No. 5 (May 1992), p. 276.

[25] Ibid.

[26] Withdrawals are a measure of the total water taken from sources (rivers and groundwater) for irrigation. Consumptive use is a measure of the water that is withdrawn that is lost to evaporation (from conveyance canals and fields) and transpired from or incorporated into crops. The difference between the two is termed return flow. Return flow includes filtration from canals, filtration from fields and surface runoff from fields. Part of return flow ultimately reaches the river from which it is taken or adds to groundwater while another portion runs off into desert hollows to form lakes (the water from these is lost to evaporation).

If discharge to the sea from the Amu Dar'ya and Syr Dar'ya averaged 56 km^3/yr between 1911 and 1960, in the 1980s it fell to 6 km^3.[27] The 1990s saw a high flow cycle. Average combined discharge of the two rivers exiting the mountains averaged 104 km^3 for 1990–98, compared to a 40-year (1959–98) figure of 94 km^3/yr.[28] Estimated inflow to the sea for this period averaged 14–15 km^3. Hydrologic probability suggests that river flow from the mountain zones of flow formation for the longer-term future will be substantially less than during the 1990s, when it was at its highest since the 1950s, although it is likely to remain above what it was in the 1980s when it averaged 87 km^3. Assuming continuation of basin withdrawals typical of the 1990s, which were substantially less than for the 1980s (see Table 3, Chapter 4), and normal flow generation conditions in the mountains, a conservative but reasonable estimate of average annual future inflow to the sea is around 10 km^3.

To restore the Aral to what it was three decades ago would require increasing average discharge to the sea by around 45 km^3. This would necessitate a somewhat larger decrease in upstream withdrawals to compensate for natural losses of the net additions to flow before they reached the sea. Assuming these are at 14 per cent (based on the figures cited above of 16 km^3 natural losses from an average annual basin flow of 116 km^3), an additional 6 km^3 reduction would be necessary for a total around 50 km^3. There seems little likelihood of attaining such a reduction of upstream use (amounting to 45 per cent of 1995 withdrawals of 111 km^3) in the foreseeable future without causing economic and social havoc for the countries of the basin that are the major users of water for irrigation. Merely to stop further shrinkage of the sea would require an increase in inflow to around 25 km^3 (some two-and-a-half times the estimated long-term figure under present conditions) and necessitate reductions in upstream withdrawals of 15 km^3 (14 per cent of the 1995 level).

Indeed, many experts (including this author) hold that additional water from the Amu Dar'ya should be used to rehabilitate the deltaic ecosystems rather than being dumped into the large Aral Sea to evaporate.[29] On the other hand, one may reasonably argue that supplementary flow from the Syr Dar'ya should be used to raise the level of the now separated small northern Aral Sea, which could be accomplished with

[27] Hydrologic data collected by the author between 1984 and 1999 from a variety of sources, including the Gidroproyekt [Hydro Planning] Institute in Moscow and Glavgidromet in Tashkent and used to derive long-term time series inflows into the Aral Sea from the Amu Dar'ya and Syr Dar'ya and water balances for the Aral Sea.

[28] Four of the years were among the highest flow for the 40-year period: 1998 (118 km^3) was second, 1993 (119 km^3) third, 1992 (118 km^3) fourth and 1993 (111 km^3) sixth; 1990 (99 km^3) and 1991 (97 km^3) were slightly above average whereas only 1997 (86 km^3) was below (based on calculations from data acquired from Glavgidromet and from Iliya Zholdaova, 'Fish Population as an Ecosystem Component and Economic Object in the Aral Sea Basin', in M. Glantz (ed.), *Creeping Environmental Problems and Sustainable Development in the Aral Sea Basin* (Cambridge: Cambridge University Press, 1999), p. 205, Table 10.1.

[29] Philip P. Micklin, 'The Aral Sea Problem', pp. 120–21.

relatively modest additions.[30] The logic of the differing approaches is based on several considerations. Even to stabilize the large Aral Sea would require increases in the flow of the Amu Dar'ya that are simply unattainable in the near-term or even medium-term future. Even if the large Aral Sea could be stabilized at its present level, its salinity is so high that it has no significant ecological or economic value.[31]

The opposite is true for the small Aral Sea (see Table 2). Because its area is much smaller than the large sea, within five years an inflow of 4.5 km^3 on an average annual basis (for 1990–98 the figure was 4.2 km^3) could raise its level to 45 metres and expand the area to around 4,000 km^2. Once the 45-metre mark was reached, this level could be maintained with a net inflow of 2.85 km^3. This would allow1.65 km^3 of more saline water in the western part of the sea to spill over to the large Aral, thereby enhancing the pace of the freshening of the sea and creating habitat conditions over the entire small Aral for the return of indigenous fishes from the Syr Dar'ya with the concomitant restoration of a commercial fishery.

In the early 1990s, local Kazakh authorities (of Aral'sk *rayon* in Kzyl-Ordinsk *oblast'*) constructed a crude dike to block the channel that had formed between the two seas. However, the dike was periodically washed away in the ensuing years. In 1997, they replaced this makeshift dike with a 20-km long, 26-metre wide dam that raised water levels 3 metres by early 1999.[32] The World Bank has approved funding for a project to construct a more technically sophisticated and stable dike, along with a control gate and channel from the western part of the small sea to the large sea.[33] The existing dam was overtopped by wave and wind-running action in April 1999 and breached with a significant loss of water.[34]

When the critical situation of the Aral Sea is taken into consideration and the amount of water needed to begin to improve conditions is added to the existing water

[30] N. V. Aladin,. and I. S. Plotnikov, 'K voprosu o vozmozhnoy konservatsii i reabilitatsii malogo Aral'skogo morya' [On the question of the possible conservation and rehabilitation of the small Aral Sea] in *Biologicheskiye i prirodovedcheskiye problemy Aralskogo morya i pryaral'ya* [Biological and natural science problems of the Aral Sea and near Aral region] Trudy Zoologicheskogo Instituta, RAN [Transactions of the Zoological Institute of the Rusian Academy of Sciences], Vol. 262, pp. 3–16; WARMAP project, *Formulation and Analysis*, pp. 66–7. If Syr Dar'ya water that in recent years has been periodically dumped into the desert Lake Arnasay (because the channel downstream of the Chardara reservoir is incapable of handling heavy flows, particularly in winter when it is ice-choked) were to reach the small Aral, it alone would substantially raise the level of this lake.

[31] However, once the large sea falls several more metres, there is the possibility of separating the deep western part from the shallow eastern part by a dike, directing the flow of the Amu Dar'ya into the western part and, over time, freshening and ecologically restoring it by allowing a controlled flow of saline water to the eastern portion (which would, nevertheless, rapidly shrink and salinize).

[32] 'Optimism rises, with water, in bid to revive Aral Sea', *Christian Science Monitor,* 5 February 1999, p. 7.

[33] World Bank, *Aral Sea Basin Program*, p. 10.

[34] Conversation with Victor Dukhovnyy, director of the Scientific Information Centre of the Interstate Coordinating Water Management Commission (SIC, ICWC), Delft, Netherlands, 19 July 1999.

uses discussed earlier, it is manifestly clear that the Aral Sea basin's water resources are stretched to the limit and beyond. Even during the high flow period of the early 1990s, after upstream withdrawals, there was insufficient inflow to the Aral to stop the sea's recession. In dry years during the 1980s, practically no water reached the sea and the desiccation proceeded at an alarmingly rapid rate.

In the next chapter we examine the most important water use by far in the Aral Sea basin – irrigation. Improvements in the region's water situation can only come about by significantly reducing the irrigated area or greatly improving the efficiency of irrigation, or by some combination of these to free substantial amounts of water.

4 IRRIGATED AGRICULTURE: THE KEY TO WATER MANAGEMENT IMPROVEMENT

4.1 The pre-Soviet period

The time-worn phrase among peoples of the world's arid regions, 'without water there is no life', is a truism.[1] Irrigation has been the mainspring of settlement and civilization in the Aral Sea region for millennia. Archaeological evidence indicates the first settlements on the piedmont plains adjacent to the Kopetdag mountains, in what is now southern Turkmenistan, arose 8–10,000 years ago.[2] Farmers cultivated wheat and barley, which provided a harvest of nearly 12,000 centners (one centner is 0.1 of a metric tonne), enough to support 6,000 people. Soviet experts contended that these crops were irrigated from small rivers exiting the mountains to the south; if so, this would be one of the earliest examples of irrigation in the world. However, recent work by Western specialists at these sites casts doubt on this claim.[3] The climate at the time was more moist than today and may have provided sufficient rainfall to grow crops without irrigation.

Nevertheless, it is clear that by the Bronze Age (the second millennium BC) farmers practised irrigation in the oases along the middle and lower course of the Amu Dar'ya and in the deltas of several terminal rivers in the Aral Sea basin (Murgab in Turkmenistan and Zeravshan in Uzbekistan).[4] Farmers raised grapes and beans as well as wheat and barley. They dug distributory canals with widths of up to 10 metres, depths of 2–3 metres and discharge capacities of 200–300 m^3/day. The irrigated area

[1] M. K. Lunezheva, A. K. Kiyatkin, and V. P. Polishchuk, '*Srednaya Aziya i Kazakhstan – drevnyaya oblast' polivnogo zemledeliya*' [Central Asia and Kazakhstan – an ancient region of irrigated cultivation] *Gidrotekhnika i melioratsiya* [Hydrotechnology and land improvement], No. 10 (1987), p. 65.

[2] V. I. Kostyukovskiy, 'Dinamika prirodnykh kompleksov ravnin Turkmenistana v usloviyakh yestestvennykh i antropogenennykh izmenenii uvlazhnennosti' [Dynamic of environmental complexes on the Turkmenistan plains in conditions of natural and anthropogenic changes in moistness], *Vodnyye resursy*, No. 4, 1992, p. 87; A. Karimov, *Upravleniye defitsitom vodnykh resursov* [Managing water resource scarcity] (Tashkent: GFNTI, 1997) p. 7.

[3] Conversations with Dr S. O'Hara, Department of Geography, University of Nottingham, 25 January 2000.

[4] Kostyukovskiy, 'Dinamika prirodnykh', pp. 87–8; A. S. Kes', B. P. Andrianov and M. A. Itina, 'Dinamika gidrograficheskoy seti i izmeneniya urovnya Aral'skogo morya' [The dynamics of hydrographic networks and changes in the level of the Aral Sea], in *Kolebaniya uvlazhnennosti Aralo-Kaspiyskogo regiona v golotsene* [Fluctuation of the moistness of the Aral-Caspian region in the Holocene], (Moscow: Nauka, 1980), pp. 188–90.

may have reached 3,500 ha, with water withdrawals of 12–13 million m^3. This implies a water withdrawal of 3,400–3,700 m^3/ha, an efficient use of water by today's standards. The population of these settlements may have reached 25–30,000. Irrigation in Khorezm, at the head of the Amu Dar'ya delta, first arose during the later part of this period.

During the first millennium BC (the Iron Age), irrigation flourished along the Amu Dar'ya, Zeravshan and Murgab.[5] The Khorezm oasis may have had 1.2 million ha under irrigation and a population of 200,000. Farmers of the Merv oasis in the delta of the Murgab irrigated some 500,000 ha which supported a population of 300,000. The total area with irrigation systems in the Aral Sea basin was huge, perhaps covering 3.5–3.8 million hectares, about half the contemporary irrigated area. However, less than half of this area was used in any given year. In the chief oases, such as Khorezm, irrigation technology advanced rapidly. Large canals were constructed directly from the main channel of the river in earthen beds that stretched several hundred kilometres, along with protective dikes to keep the river from overflowing its banks and flooding crops. The first evidence of irrigation on the lower Syr Dar'ya dates to the middle of the first millennium BC – a thousand years later than on the Amu Dar'ya.

However, around the 4th and 5th centuries AD, the major irrigation systems of Khorezm were destroyed during a social and economic crisis.[6] Irrigation was restored by the 7th century but was partially destroyed during the Arab invasion of 712. Irrigation again flourished during the 12th and at the beginning of the 13th centuries, when Khorezm became the centre of a new feudal state whose government saw to the restoration of the canals and protective dikes. Large tracts of new land were irrigated and major canals built far into the Kyzyl-Kum desert to unite the irrigated zones of the Amu Dar'ya and Syr Dar'ya. The irrigation experts of the time perfected the method of using diversion dams and storage basins to capture floodwaters for irrigation. During this period the land with irrigation facilities within the area controlled by Khorezm was near 2.5 million ha. Because of improved technology, such as the use of the chigir wheel, farmers were able to irrigate more efficiently and continuously than previously.[7]

The Mongol invasion of Central Asia in 1220 led to the nearly complete destruction of irrigation systems along the Amu Dar'ya and Syr Dar'ya.[8] Dikes and

[5] Ibid.

[6] Kostyukovskiy, 'Dinamika prirodnykh', pp. 88–9; Kes' et al., 'Dinamika gidrograficheskoy', pp. 191–2.

[7] The chigir is a wooden wheel, some 2 metres in diameter, with clay buckets arranged around the rim. The wheel is mounted on a horizontal axis in a pit of water. The wheel can be powered by draft animals (horse, donkey, camel) or the force of water to lift water from the pit to the fields. The device was widely used from ancient times until the Soviet transformation of agriculture in Central Asia during the late 1920s and 1930s (Karimov, *Upravleniye defitsitom*, p. 13).

[8] Kes' et al., 'Dinamika gidrograficheskoy', pp. 192–94; Kostyukovskiy, 'Dinamika prirodnykh', p. 89.

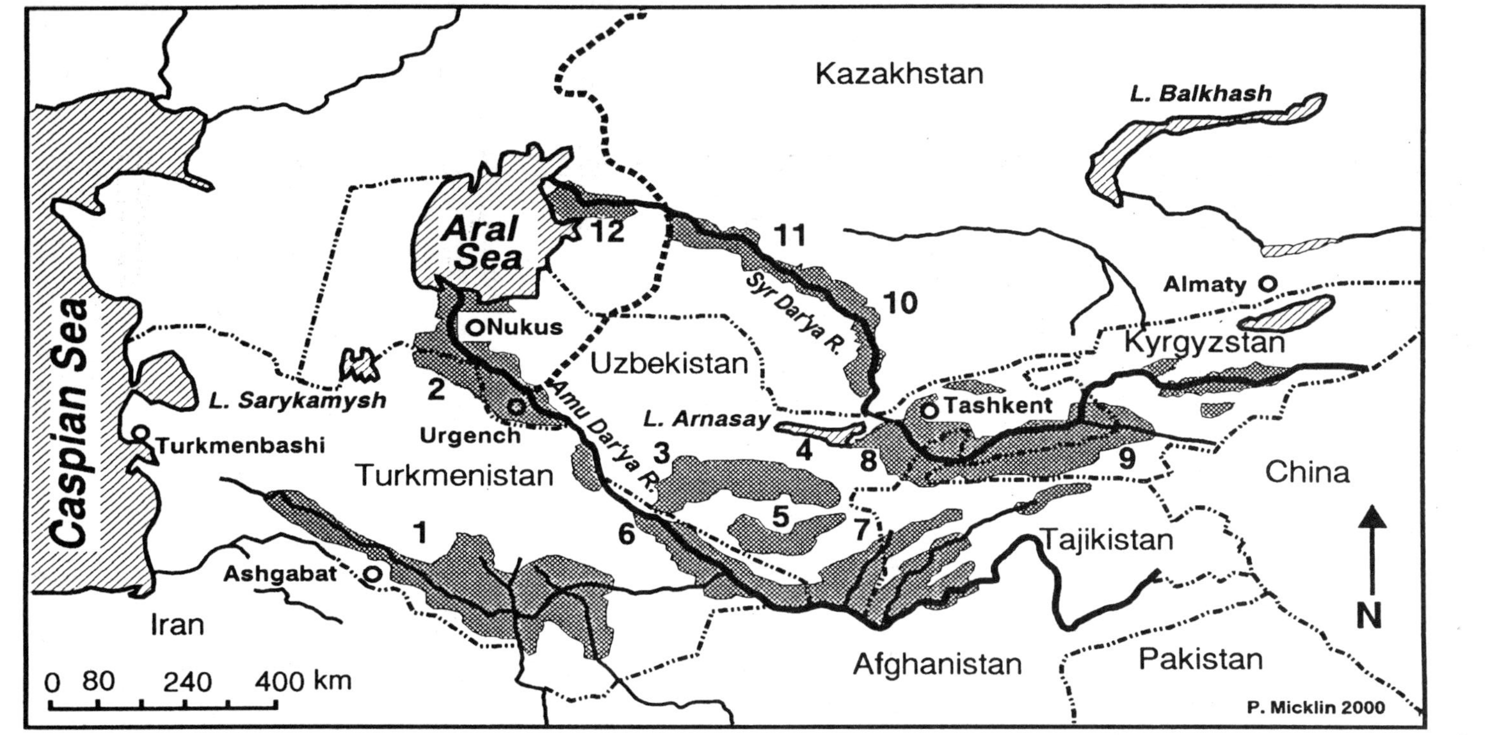

MAJOR IRRIGATION COMPLEXES IN THE ARAL SEA BASIN

1. Kara-Kum Canal
2. Amu Dar'ya Delta
3. Amu-Bukhara Canal
4. Zeravshan Valley
5. Karshi Steppe
6. Middle Amu Dar'ya
7. Surkhandar'ya Valley
8. Golodnaya Steppe
9. Fergana Valley
10. Middle Syr Dar'ya
11. Kzyl-Orda Canal
12. Syr Dar'ya Delta

main irrigation zones in the Aral Sea Basin

Proposed Siberia-Aral Sea Canal

Map 2: Irrigation development in the Aral Sea basin

dams that controlled the waters of the rivers and allowed their diversion for irrigation were wrecked. The Amu Dar'ya diverted itself westward to the Sarykamysh depression (as it had several times before), depriving the lower delta and Aral Sea of water.[9] Irrigation facilities were slowly rebuilt and the Amu Dar'ya restored to its former course. By the early 19th century the irrigated area was around 2 million ha, less than it had been prior to the Mongol attack.

Between the 1860s and 1900 the Russian empire brought Central Asia, including the Aral Sea basin, under its control, mainly by conquest.[10] The region was known as Russian Turkestan or just Turkestan. A major expansion of irrigation, begun around 1900, focused on construction of new systems within the bounds of the traditional irrigated areas (Tashkent, Bukhara, Khorezm, Fergana and Zeravshan valleys) as well as expansion into desert areas that had never been irrigated, chiefly the Golodnaya (Hungry) steppe (see Map 2). By 1913, the irrigated area reached 3.2 million ha (see Table 3). The newly irrigated lands allowed an expansion of the area sown to cotton to 556,000 ha (17 per cent of the total irrigated zone). The major reason for this was to meet the needs of the growing textile industry in Russia.[11] However, cotton also displaced traditional food crops in portions of the old irrigated zones. For the Aral Sea basin, yields of cotton averaged 12 centners/ha.

Irrigation expansion in the early part of the 20th century occurred mainly in the alluvial fans, where rivers exited the mountains, and downstream in the river valleys and deltas.[12] This allowed development of the most fertile parts of the piedmont plain and dry sections of river deltas using gravity diversions that did not require construction of head-works facilities (dams, pumping stations, etc.). The chief task was

[9] Diversion of the Amu Dar'ya westward so that it flowed into the Sarykamysh hollow (and sometimes farther through the Uzboy channel to the Caspian Sea after it overtopped Sarykamysh) rather than the Aral Sea has occurred periodically during the Holocene geological epoch (the last 10,000 years). These diversions have been caused both by natural events (sedimentation of the bed and subsequent breaching of the river's left bank during the spring floods) and by human actions (destruction of dikes and dams built to keep the river flowing to the Aral). Until the 1960s, these diversions were the primary cause for the fluctuations of the Aral Sea's level. See Kes' et al., 'Dinamika gidrograficheskoy', pp. 185–97, and A. S. Kes'and I. A. Klyukanova, 'O prichinakh kolebaniya urovnya Aral'skogo moray v proshlom' [On the reasons for fluctuations in the level of the Aral Sea in the past], *Izvestiya Akademii Nauk SSSR, seriya geograficheskaya*, No. 1, 1990, pp. 78–86.

[10] A. Brown, M. Kaser and G. Smith (eds), *The Cambridge Encyclopedia of Russia and the Former Soviet Union* (Cambridge: Cambridge University Press, 1994), pp. 536–8; R. S. Clem, 'The Frontier and Colonialism in Russian and Soviet Central Asia', in R. A. Lewis (ed.), *Geographic Perspectives on Soviet Central Asia* (New York: Routledge, 1992), pp. 26–33.

[11] R. S. Clem, 'The Frontier', p. 32.

[12] E. Karimov, 'Chelovek i priroda: pora zaklyuchat' soyuz' [Man and nature: time to enter an alliance] *Ekonomika i statistika* [Economics and statistics], No. 3, 1995, p. 48–9; Ye. I. Pankova, I. P. Aydarov, I. A. Yamnova, A. F. Novikova and N. S. Blarovolin, *Prirodnoye i antropogennoye zasoleniye pochv basseyna Aral'skogo morya (geografiya, genisis, evolyutsiya)* [Natural and anthropogenic salinization of the Aral Sea basin soils (geography, genesis, evolution)] (Moscow: Rossiyskaya Akademiya Sel'skokhozyaystvennykh Nauk [Russian Academy of Agriculture], 1996), pp. 76–7.

Table 3: Irrigation data for the Aral Sea basin (1913–95)[a]

Year	Irrigated area (million ha)	Withdrawals for irrigation (km³)	Withdrawals for irrigation (m³/ha)[b]	Cotton area million (ha)	Cotton as % of total irrigated area	Cotton yields centners/ ha)	Water for raw cotton (m³/centner)
1913	3.2	25.6–43.2	8000–13,500	0.556	17.4	12	950
1922	1.7	16	9400	0.100	5.9	7	1230
1933	3.5	40	11,500	1.800	51.4	5	1800
1940	3.8	49	13,000	1.369	36.0	14	1800
1945	no data	no data	15,000	1.110	no data	10	1350
1950	3.8	57	15,000	1.580	41.6	20	800
1965	4.8	82	17,000	2.287	47.6	23	800
1980	6.3	107.1–126	17,000–20,000	2.869	45.5	29	750
1985	7.0	112–133	16,000–19,000	3.051	43.6	26	700
1990	7.25	109	14,600–17,000	2.909	40.1	27	800
1995[c]	7.94	100	12,594	2.574	32.4	24	200

Sources: Taken or calculated from data in Ye. I. Pankova, I. P. Aydarov, I. A. Yamnova, A. F. Novikova and N. S. Blarovolin, *Prirodnoye i antropogennoye zasoleniye pochv basseyna Aral'skogo morya (geografiya, genesis, evolyutsiya)* [Natural and anthropogenic salinization of the Aral Sea Basin soils (geography, genesis, evolution) (Moscow: Rossiyskaya Akademiya Sel'skokhozyaystvennykh Nauk, 1996), pp. 74–85, Figures 11, 12, 14 and 15; *Strana Sovetov za 50 let* [Land of the Soviets for 50 years], sbornik statisticheskikh materialov [A collection of statistical material] (Moscow: Statistika, 1967, pp. 140; *Narodnoye khozyaystvo SSSR za 60 let* [National economy of the USSR for 60 years] (Moscow: Statistika, 1977, p. 310; *Narodnoye khozyaystvo SSSR, 1922–1982* [National economy of the USSR, 1922-1982] (Moscow: Finansy i Statistika, 1982, p. 251; *Narodnoye khozyaystvo SSSR v 1973g.* [National economy of the USSR in 1973] (Moscow: Statistika, 1974), pp. 402–3; *Narodnoye khozyaystvo SSSR v 1990g.* (Moscow: Finansy i Statistika, 1991), pp. 472–3; K.Sh. Siprozhidnikov, 'O vyyavlennykh prichinakh snizheniya urovnya Aral'skogo morya', *Problemy osvoyeniya pustyn'*, No. 6, 1991, pp. 24–5; V. Dukhovnyy, *Melioratsiya i vodnogo khozyaystva aridnogo zona* [Reclamation and water management of the arid zone] (Tashkent: Mekhnat, 1993), pp. 50, 56, 63; Philip P. Micklin, *The Water Management Crisis in Soviet Central Asia*, The Carl Beck Papers in Russian and East European Studies, No. 905 (Pittsburgh, PA: The Center for Russian and East European Studies, August 1991), pp. 16–19; World Bank and the ICWC, *Developing a Regional Water Management Strategy: Issues and Work Plan*, Aral Sea Basin Program Technical Paper Series, Washington, DC, April 1996, pp. 20–22; ICAS, *Fundamental Provisions of Water Management in the Aral Sea Basin: A Common Strategy of Water Allocation, Rational Water Use and Protection of Water Resources*, prepared with the assistance of the World Bank, October 1996, Chapter 10, Tables 10.1, 10.2; UNESCO Division of Water Sciences and Scientific Advisory Board for the Aral Sea Basin, *Water Related Aral Sea Vision for the Year 2000*, Draft of 31 October 1999, p. 64.

[a] Excludes Afghanistan and Iran.

[b] For some years, data from different sources had significant variations; these are given as ranges rather than a single figure.

[c] Cotton data are for 1993 and 1994.

the building of main delivery canals. In the drier parts of the Amu Dar'ya delta, farmers used a system of deepened earthen canals with chigir wheels to lift the water to the fields. These canals simultaneously fulfilled the role of distribution network for irrigation water and collector network for drainage water. Initially, these efforts went well and substantially increased production. But after 10–15 years, the first signs of waterlogging and secondary soil salinization (from over-irrigation, lack of proper drainage and irrigation of naturally saline soils) appeared in the Golodnaya steppe, leading to the loss of 60 per cent of the newly irrigated area.

4.2 The Soviet period

During the Civil War (1919–24) following the Bolshevik seizure of power, irrigation systems were destroyed or left idle.[13] The irrigated area decreased from 3.2 million ha in 1913 to 1.7 million ha in 1922. The area on which cotton was grown shrank to a little over 100,000 ha and average yields fell to near 7 centners/ha. With the consolidation of Soviet power in Central Asia, Moscow ordered the restoration and repair of irrigation systems, which was completed between 1925 and 1928. At this time, although there were significant variations from one place to another, the crop structure for the Aral Sea basin in terms of the total area sown was 30–40 per cent grains, 30 per cent perennial grasses, 5–25 per cent rice and no more than 20–25 per cent cotton.[14]

The traditional system of irrigated agriculture in the Aral Sea basin was essentially preserved until the mid-1920s. As indicated above, this system had been developed over thousands of years and had survived wars, social collapse, natural catastrophes and the Tsarist conquest of the region. It had both a land use and an administrative component. In terms of land use, some 90 per cent of the irrigated area consisted of small, individually managed farms of 2–3 hectares.[15] These farms were typically run by extended families; some owned the land and some rented the land or worked it as tenants of large landowners for a share of the crop. The farms were divided into irrigated fields of 0.3–0.8 ha with permanent low walls that were lined with trees. Farmers used primitive irrigation technologies based on manual labour supplemented by draught animals. Animal wastes were used as fertilizer. Filtration losses in the irrigation system were as high as 30 per cent. Irrigation systems were not built according to any overall plan and they lacked control structures, drainage canals and discharge canals to convey run-off from fields.

[13] Ibid.

[14] Pankova et al., *Prirodnoye i antropogennoye*, p. 78.

[15] K. Sh. Siprozhidnikov, 'O vyyavlennykh prichinakh snizheniya urovnya Aral'skogo morya' [On the apparent reasons for the drop in the level of the Aral Sea] *Problemy osvoyeniya pustyn'*, No. 6, 1991, pp. 24–5.

We might expect such systems to be unsustainable as a result of rapid waterlogging, soil salinization and falling crop yields. But they obviously were not. The principles of irrigated farming developed by trial and error over the centuries were sound. The small size of the flooded fields ensured even distribution of the water and the absence of any wasteful run-off. The banks which surrounded these sectors acted to absorb and accumulate salt, preventing soil salinization. The trees planted along the banks transpired excess water to control waterlogging as well as lowering wind speeds over the fields, which reduced evaporative losses from the soil. The trees also served as a fuel source. Water was used efficiently. Average withdrawals, including filtration losses on the fields and in delivery canals, are estimated to have been 10,700–11,500 m^3/ha.

The administrative governance system of irrigation had also evolved over a lengthy period. The basic principle was central governmental control of water management, but local responsibility for operation and maintenance of the water management system.[16] Islamic law (*Sharia*) saw water as social property and did not allow private ownership of it. Local authorities assigned water use rights to farmers and ensured the maintenance of irrigation facilities for distributing water to them. Higher levels of government ensured that local officials fulfilled their responsibilities and handled disputes between regions (e.g. Samarkand and Bukhara on the Zeravshan river) or those that could not be resolved by local leaders.

The government required water users to both build and maintain the irrigation system. This was managed through a complicated system of obligatory annual work to clean canals of sediments, repair dams and other facilities and deal with any other maintenance or construction issues that needed attention. The authorities also collected taxes for water used in irrigation and fines from those who did not appear for their obligatory duties. They used the collected funds to finance the construction and maintenance of the irrigation systems.

The irrigation water management system operated through a hierarchy of water technicians/bureaucrats.[17] For example, during the 19th century in Bukhara khanate the chief vizier of the state (known as the *atalyk*) was in charge of irrigation. The *mirab* (manager) of the main irrigation canal was subordinate to him and responsible for water distribution from and maintenance of this facility. The *mirabs* of the main distributory canals were next in the hierarchy and had the same responsibilities for the facilities under their control. The *mirabs* of the canals for distribution of water to individual fields were at the bottom of the management system. There were also observers (*panzhbegi* and *obron*) to ensure that the off-takes at critical diversion

[16] A. Karimov, *Upravleniye defitsitom*, pp. 8–11; A. Karimov, 'Water regimes in Central Asia', paper given at the Aral Sea Basin Workshop, Tashkent, Uzbekistan, May 19–21, 1998 (conference papers are available from the sponsor – the Social Science Research Council in New York).
[17] Ibid.

points along the distributory canals conformed to the limits set by higher authorities, and personnel known as *obandozy* who were assigned responsibility to make sure the orders of the *atalyk* were delivered to the *mirabs* of the main canals. This system made it possible to determine who was responsible for problems in the system and to punish the guilty (sometimes severely).

The traditional system of irrigated agriculture was adjusted to the environmental constraints of an arid environment and provided strong incentives for farming families to cooperate in the maintenance of irrigation systems as well as to use both land and water carefully.[18] The Soviet authorities could have kept the best aspects of this system while carefully and judiciously introducing more modern Western technology, where appropriate, and pursuing a programme of land reform and redistribution to give land to the landless peasantry. Instead, they chose to institute historically unprecedented changes in irrigated agriculture in the Aral Sea basin that radically and permanently changed the face of agriculture and peasant life.[19]

There were two primary motivations: first, there were deeply held beliefs among the Bolshevik leadership and intelligentsia that the traditional (in their view feudal) irrigation and agricultural system in Central Asia was primitive, backward and inefficient, as well as exploitative and oppressive of the peasantry that formed its basis; second, there was the need to institute state control of irrigation and agriculture, as well as individual farming, in order to dramatically expand the production of cotton.[20] In the view of the Bolsheviks this 'corrupt and worthless' system had to be entirely destroyed and replaced by centrally controlled, modern, large-scale, equipment-intensive irrigation on the industrial model with tight government control where the peasantry, rather than being owners/managers of the land and water, were simply employees executing the state economic plan.

The first steps, instituted by decree in 1925, nationalized all lands and began the confiscation of large landholdings for redistribution to small landholders and landless peasants.[21] Nevertheless most of the irrigated area remained under private management. Over the next few years, the Soviet government instituted other major changes. Individual farms were amalgamated into larger, supposedly more efficient units.[22] As

[18] Siprozhidnikov, 'O vyyavlennykh prichinakh', pp. 24–5; A. Karimov, 'Water regimes'.

[19] An excellent and detailed treatment of the devastating changes wrought on farmers and agriculture in Uzbekistan under the Soviet system is provided in J. Thurman, *The 'Command-Administrative System' in Cotton Farming in Uzbekistan 1920s to the Present*, Papers on Inner Asia, No. 32 (Bloomington, IN: Indiana University Research Institute for Inner Asian Studies, 1999), pp. 2–34.

[20] Askochenskiy, *Orosheniye i obvodneniye v SSSR*, pp. 13–20; Siprozhidnikov, 'O vyyavlennykh prichinakh', p. 25.

[21] E. Karimov, 'Chelovek i priroda ...', p. 48. In Uzbekistan, which accounted for the largest share of the irrigated area in the Aral Sea basin, large landholders controlled more than half the irrigated territory.

[22] Siprozhidnikov, 'O vyyavlennykh prichinakh,' p. 25.

a part of this process, and contrary to promises made, the small irrigation fields of 0.3–0.8 ha were combined into larger units, averaging 3.5 ha, resulting in the destruction of the earthen walls separating them and the uprooting of the trees growing on them. The area of irrigation grew, mainly by conversion of lands formerly used for growing trees or kept in fallow within the bounds of irrigation systems. As a result, waterlogging and soil salinization increased significantly. The area devoted to cotton grew rapidly, but yields were far below pre-revolutionary levels. Average irrigation water withdrawals in Uzbekistan rose from 10,000 to 13,000 m^3/ha. Water waste and losses in irrigation increased sharply.

As in other parts of the USSR, the process of massive collectivization of agriculture in Central Asia began in 1929. Soviet authorities forced individual farms to combine into *kolkhozy* (collective farms) while *sovkhozy* (state farms) were set up on newly irrigated lands. Farmers' use of distinct, separate portions of the land was abolished and people were organized into teams, working the land in common. The main goal was to reach cotton independence by the end of the first five-year plan in 1933.[23] Cotton planting grew to over 50 per cent of the sown area and domestic production of cotton rose from 59 to 97 per cent of national needs. However, yields remained below 8 centners/ha while water use rocketed to 1,800 m^3/centner – it had been around 1,200 m^3/centner in the early mid-1920s (see Table 3).

The situation improved during the remaining years of the decade as the authorities eased the more onerous aspects of collectivization and slowed the pace of its implementation. Renovation of irrigation systems and mechanization of fieldwork slightly lowered water use. By 1940 the irrigated area rose to 3.8 million ha but cotton's percentage of that total fell to 36 per cent. Average cotton yields in the Aral Sea basin nearly tripled.

By the end of the decade collectivization was firmly in place. The lasting legacy of this social and economic transformation was the destruction of traditional irrigation in the Aral Sea basin. Irrigated agriculture was converted into a completely state-controlled enterprise and set the stage for subsequent developments in irrigation for the rest of the Soviet period.

Even though Central Asia was not invaded, the war years (1940–45) saw a deterioration of irrigated agriculture.[24] The irrigated area shrank, cotton plantings dropped by nearly 20 per cent and yields of cotton fell to 10 centners/ha as water use per centner declined sharply. Irrigation more than recovered by 1950 as the irrigated area regained pre-war levels, cotton hectarage increased to 1.6 million ha (42 per cent of the irrigated area), cotton yields doubled to 20 centners/ha and water use per centner of cotton dropped to 800 m^3.

[23] Pankova et al., *Prirodnoye i antropogennoye*, p. 78–81; E. Karimov, 'Chelovek i priroda', p. 49.
[24] Pankova et al., *Prirodnoye i antropogennoye*, p. 74; E. Karimov, 'Chelovek i priroda', p. 49.

Irrigation in the Aral Sea basin underwent steady expansion from 1950 to 1965, reaching nearly 5 million ha (see Table 3).[25] The Soviet government placed emphasis on expanding irrigation in older irrigated areas such as the Bukhhara oasis, the Fergana valley, and the lower Amu Dar'ya and Syr Dar'ya as well as in the zones of more recent irrigation (e.g. Golodnaya and Karshi steppes of Uzbekistan; see Map 2). Construction on the Kara-Kum canal, the largest capacity and longest irrigation canal in the former USSR, started in 1954. Irrigation during this period expanded onto the lower-lying portions of the piedmont plains, which have poor drainage and are natural accumulators of salt flowing in groundwater from higher territory. This exacerbated the already severe problems of waterlogging and soil salinization.

What to do with increasing amounts of saline irrigation drainage water became a serious question. Much of it was simply dumped into rivers, worsening their water quality. Drainage water also began to flow into desert depressions, forming permanent lakes. Today, the largest of these are Sarykamysh in Turkmenistan and Arnasay in Uzbekistan, each of which has grown to several thousand km^2 in area (see Map 2). From 1950 to 1965, irrigation water withdrawals in Uzbekistan, the republic with by far the largest irrigated area, rose by a factor of 1.25 while irrigation drainage flows grew more than threefold.

Construction of very large irrigation systems encompassing hundreds of thousands of hectares (e.g. Golodnaya and Karshi steppes) and extending well into the deserts started during these years. These systems were built on an industrial basis and included social amenities and infrastructure (houses, stores and entertainment and recreational facilities) and communication networks (roads, telephones lines). To facilitate the use of larger tractors and other equipment, the field size was dramatically increased to as much as 100 ha.[26] These new systems were supposed to improve irrigation conditions and performance. However, they did the opposite. The huge fields were impossible to irrigate evenly, requiring application of excessive quantities of water. This led in turn to soil erosion, rising groundwater levels and waterlogging, secondary salinization, increased water use and lowered yields. Furthermore, some of the areas had naturally saline soils, which required annual applications of water to leach the salts prior to the start of the irrigation season. Another major problem was that the push to expand the irrigated area took precedence over the proper construction of irrigation systems. Construction brigades frequently disregarded the requirement to install drainage facilities in the drive to maximize the area of new irrigation.

The cotton area grew to 2.3 million ha or 48 per cent of the irrigated area by 1965 (see Table 3).[27] Yields slowly increased to 23 centners/ha. During this period,

[25] E. Karimov, 'Chelovek i priroda', pp. 49–50; Pankova et al., *Prirodnoye i antropogennoye*, pp. 76–7.
[26] Siprozhidnikov, 'O vyyavlennykh prichinakh', pp. 25–7.
[27] Pankova et al., *Prirodnoye i antropogennoye*, p. 76–85.

fertilizer use for cotton rose rapidly. Average water withdrawals in the Aral Sea basin rose to around 17,000 m³/ha, with about 11,000 m³/ha (64 per cent) lost, i.e. not being used for plant growth needs. Similar figures for the mid-1920s, prior to the Soviet transformation of irrigated agriculture, are around 10,000 and 5,000 m³/ha respectively.

The next 20 years, 1965–85, saw the rapid development of irrigation in the Aral Sea basin, with the irrigated area approaching 7 million ha by the end of the period.[28] The main emphasis was on expanding irrigation in newly irrigated zones, including the Golodnaya, Djizak, Karshi and Kashkadar'ya steppes, and along the Amu-Bukhara canal (which was under construction) in Uzbekistan and in the Yavansk valley in Tajikistan (see Map 2). The extension of the Kara-Kum canal in Turkmenistan also continued.

This was the era of construction of large (and very large) dams and reservoirs along the Amu Dar'ya, Syr Dar'ya and their tributaries.[29] The largest of these are the Toktogul, Andizhan, Kayrakkum, Charvak and Chardar'ya along the Amu Dar'ya and its main tributaries, the Naryn, Karadar'ya and Chirchik, and the Tyuyamyuyunsk and Nurek on the Chirchik and its chief tributary, the Vakhsh. The fundamental purpose was flow regulation to provide more water for irrigation (by storing spring high flows for use during summer) and provision of hydropower, much of which would be used for powering irrigation pumps. By the 1980s, the flow of the Syr Dar'ya was nearly completely regulated and that of the Amu Dar'ya was largely controlled.

The Soviet government devoted much attention to installing collector-drainage networks to cope with the growing problems of rising groundwater levels, waterlogging and secondary salinization. However, this increased return drainage flows into rivers, substantially worsening their quality. Another major goal was raising the efficiency of irrigation systems. Until 1980, withdrawals increased, reaching a peak somewhere between 17,000 and 20,000 m³/ha, but by 1985 they fell to around 16,000–19,000 (see Table 3, footnote b). Water losses in irrigation systems peaked in the 1970s at 11–12,000 m³/ha and fell to around 10,000 by 1985.[30] In an attempt to maintain cotton yields, the use of mineral fertilizers and pesticides grew enormously, also contributing to the pollution of rivers by return flows from irrigated fields. Cotton's percentage of the irrigated area fell slightly to 44 per cent by 1985. Yields rose to a peak of 29 centners/ha in 1980 (in the 1930s they had been as low as 5

[28] Ibid.; E. Karimov, 'Chelovek i priroda', p. 50. The period 1980–85 was characterized by very rapid growth as irrigators attempted to maintain cotton harvests in the face of declining yields and also expand the irrigated area devoted to grains.

[29] Pankova et al., *Prirodnoye i antropogennoye*, p. 76; V. Dukhovnyy, *Melioratsiya i vodnoye khozyaystvo aridnogo zona* [Reclamation and water management of the arid zone] (Tashkent: Mekhnat, 1993), pp. 260–61; Philip P. Micklin, *The Water Management Crisis in Soviet Central Asia*, The Carl Beck Papers in Russian and East European Studies, No. 905 (Pittsburgh, PA: The Center for Russian and East European Studies, August 1991), pp. 30–32.

[30] Pankova et al., *Prirodnoye i antropogennoye*, p. 84, Figure 14.

centners/ha) but dropped to 26 centners/ha by 1985. (However, the late 1970s and early 1980s were the time of the infamous cotton scandals, when harvest and yield data were deliberately exaggerated; hence the 1980 figure may be exaggerated by as much as 20 per cent – see Chapter 6).[31]

With the rapid growth of withdrawals for irrigation, the water resources of the Amu Dar'ya and Syr Dar'ya were approaching exhaustion. The level of the Aral Sea was falling. Central Asian political leaders and national and regional water managers began in the late 1960s to push hard for the implementation of grandiose plans to divert huge quantities of water from Siberian rivers (Ob' and Irtysh) to Central Asia (see Map 2).[32] The major purpose of this was to provide water to continue expanding irrigation, not to replenish the drying Aral. Experts completed the design of this project by the late 1970s. Although there was considerable opposition to the endeavour from nationalist writers, environmentalists and scientists in Russia, it appeared to be on the verge of implementation by the early 1980s. However, after Gorbachev assumed the leadership in 1985, these plans were vehemently attacked as environmentally dangerous and economically unjustified. The project was put on permanent hold in 1986. Moscow told Central Asians that they would and could get along with the water available within the region. Central Asian political leaders and water managers are still bitter about the cancellation of the Siberian project, which they believe was promised them by Moscow in exchange for growing ever more cotton.

The long expansion of irrigation in the Aral Sea basin slowed considerably during the last years of the Soviet regime (1985–91).[33] The formerly powerful and rich water management establishment was composed of the national Ministry of Water Management in Moscow and its republican affiliates. For decades it had successfully lobbied for more irrigation, but now it lost its influence. With the open approval of the new Gorbachev regime the ministry was publicly, and often bitterly, attacked for squandering state funds on economically and ecologically questionable water development projects and for destroying the Aral Sea through the massive expansion of irrigation in Central Asia. The fresh-water resources of the region, given current irrigation practices, were exhausted and there would be no rescue from Siberian rivers. It was time to cease initiation of major new projects and concentrate on improving irrigation efficiency and dealing with the very serious problems of soil salinization and waterlogging in already irrigated zones. Another stated government goal was to provide more water to the Aral Sea.

The new emphasis on irrigation improvement had some success.[34] Cotton

[31] Thurman, *The 'Command-Administrative System'*, pp. 41, 46–7.

[32] Micklin, *The Water Management Crisis*, pp. 60–68.

[33] Ibid., pp. 60–82; E. Karimov, 'Chelovek i priroda', p. 50.

[34] Pankova et al., *Prirodnoye i antropogennoye*, p. 78–84; E. Karimov, 'Chelovek i priroda', p. 50; A. Karimov, 'Water regimes'.

decreased to 40 per cent of the sown area by 1990, with average basin-wide yields rising slightly to 27 centners/ha (see Table 3). From the early 1980s, water withdrawals per hectare steadily declined. Renovation and improvement of irrigation systems played a role in this but the main factor was a series of dry years in the 1980s that forced strict limitations on water use. On the negative side, groundwater levels continued to rise, aggravating the problems of waterlogging and the process of soil salinization. By 1989, 3.6 million ha (50 per cent of the irrigated area in the Aral Sea basin) suffered from salinization. The volume of drainage return flows was still growing rapidly and accounted for over 40 per cent of irrigation withdrawals. Part of this flow re-entered rivers, causing average salinity in the lower Amu Dar'ya and Syr Dar'ya to rise to 2–3 grammes/litre by 1990.

4.3 Contemporary irrigation

The new states of Central Asia formed from the USSR after its collapse at the end of 1991 inherited an irrigation system developed over a 65-year period. This system had grown under the principles of state ownership and tight control, centralized (top-down) management which allowed little initiative at the local level, a focus on the dominant production of cotton and the need for large-scale facilities. As independent entities for the past eight years, the new states have dealt with this legacy individually. However, given their common history and the integrated nature of the irrigation system, they face similar problems in adapting to the future. This section provides a description and analysis of irrigation in the Aral Sea basin in the mid- to late 1990s, including an evaluation of the potential for solving the most serious irrigation issues faced by the basin nations.

The key to improving management of the Aral Sea basin's water resources (including the provision of substantial additional quantities of water for the Aral Sea, for the deltas of the Amu Dar'ya and Syr Dar'ya, and for expanding economic uses) is irrigated agriculture. A recent World Bank report cites the irrigated area in the basin in 1995 (excluding Afghanistan and Iran) at 7.94 million hectares (see Tables 3 & 4 and Map 2). Uzbekistan has the majority of the irrigated area, with Turkmenistan a distant second. Irrigation is by far the most important source of water withdrawals in the basin, accounting for 92 per cent of the total in that same year (see Table 5). Uzbekistan again leads the pack, trailed by Turkmenistan. Irrigation is a heavy consumer of water. Much of the water that is withdrawn is not returned to the source (river or groundwater) from which it was taken but is evaporated from fields, transpired from crops or runs off into desert depressions to be lost by evaporation. Taking this into consideration, irrigation's contribution in terms of consumption to the diminution of basin water resources is even greater than its share of withdrawals implies.

Table 4: Irrigated areas in the Aral Sea basin in 1995 (million ha)

Country	Amu Dar'ya basin	%	Syr Dar'ya basin	%	Aral Sea basin	%
Uzbekistan	2.48	53	1.80	55	4.28	54
Turkmenistan	1.74	37	0.00	0	1.74	22
Tajikistan	0.43	9	0.29	9	0.72	9
Kazakhstan	0.00	0	0.74	22	0.74	9
Kyrgyzstan	0.00	0	0.46	14	0.46	6
Total	4.65	100	3.29	100	7.94	100

Source: Adapted from World Bank, *Aral Sea Basin Program*, Table 2. Excludes Afghanistan and Iran.

Table 5: Water withdrawals for irrigation in the Aral Sea basin in 1995 (km^3)

Country	Amu Dar'ya basin	Syr Dar'ya basin	Aral Sea basin	% of Aral Sea basin	Specific withdrawal (m^3/ha)	Total basin water withdrawals	Irrigation % of total
Uzbekistan	33.2	19.8	53.0	53.0	12,383	58.0	91
Turkmenistan	22.4	0.0	22.4	22.4	12,874	23.1	97
Tajikistan	7.0	3.3	10.3	10.3	14,306	12.0	86
Kazakhstan	0.0	9.7	9.7	9.7	13,108	11.0	88
Kyrgyzstan	0.0	4.6	4.6	4.6	10,000	5.1	90
Total	62.6	37.4	100.0	100.0	12,594[a]	109.2	92

Source: Adapted from World Bank, *Aral Sea Basin Program*, Table 2. Excludes Afghanistan and Iran.
[a] Average.

Substantial savings of water are possible in the irrigation sector through contraction of the irrigated area, improvements in irrigation efficiency and switching from higher to lower water use crops. The irrigated area in the Aral Sea basin was 7.25 million ha in 1990.[35] It rose to 7.94 million ha in 1995, a 9.5 per cent increase.[36] All the basin states except Kazakhstan plan imminent increases in the irrigated area: Kyrgyzstan by over 400,000 ha; Tajikistan by between 40,000 and 140,000 ha; Turkmenistan by 600,000 ha and Uzbekistan by between 420,000 and over 600,000 ha.[37] A common argument used to justify the increases is the need to expand food production to meet

[35] Dukhovnyy, *Melioratsiya i vodnoye*, p. 56.

[36] This figure may be exaggerated, as a 10 per cent growth over 5 years, given the tight water situation in the basin, seems improbable. Growth from 1985 to 1990 was less than 4 per cent.

[37] ICAS, *Fundamental Provisions of Water Management in the Aral Sea Basin: A Common Strategy of Water Allocation, Rational Water Use and Protection of Water Resources*, prepared with the assistance of the World Bank, October 1996, Chapter 6; V. Antonov, 'O programme dal'neyshego razvitiya orosheniya v Uzbekistane' [On the programme for further development of irrigation in Uzbekistan], *Vestnik Arala*, No. 1 (vodnyye resursy), spring 1996, pp. 7–10.

population growth. Significant water savings through reduction of the irrigated area are unlikely for the foreseeable future.

Substantial water savings in irrigation, therefore, must come through improvements in efficiency and the replacement of highly water-dependent crops with lower use varieties. Arriving at savings from raised efficiency entails determining the minimum amount of water that needs to be withdrawn in a given area for optimal (or near optimal) growth of a specific crop mix; the difference between this figure and what is actually withdrawn represents potential gross savings. However, the net additions to usable water resources would be lower, as corrections are necessary to take account of reduced drainage water return flows to rivers. Calculating the minimum (or norm in the Soviet parlance still used in Central Asia) is not easy. It depends on the availability of detailed and accurate data on climate, soil conditions and crops on a scale sufficiently large to reflect the regional variability in the Aral Sea basin, as well as calculating the minimum obtainable losses in the irrigation delivery system.

An authoritative figure cited at the end of the Soviet period for the minimum average field application rate obtainable was 8,500 m^3/ha.[38] Assuming the average losses in the delivery canals could be reduced to 20 per cent, meaning that 80 per cent of the water would reach distribution points to the fields, the overall withdrawal would be 10,600 m^3/ha. Subtracting this figure from the basin-wide withdrawal for 1995 of 12,594 m^3/ha indicates possible savings of 1994 m^3/ha. This would translate into a basin-wide gross reduction of 15.8 km^3/yr. Net savings (corrected for reductions in return flows to rivers) would be smaller. Assuming return flows to rivers at 24 per cent of withdrawals, typical for the first half of the 1990s, and assuming that they would be reduced by the same percentage as withdrawals (15.8 per cent), net savings would equal 12 km^3/yr. The gross figure falls within the range of estimated feasible water savings (12.7–18.3 km^3) made by the countries of the region in the mid-1990s. The net figure is slightly less than the lower end of the range.[39]

These savings are based on modern technologies for reducing irrigation water usage such as lining delivery canals to reduce water losses and more precise levelling of fields using lasers to ensure even distribution of water. Substantially greater savings could be achieved by combining these techniques with other measures such as irrigation scheduling using computers and specialized software and real-time monitoring of soil moisture and crop water needs to determine exactly when and how much water to apply. Other measures include genetic engineering of crops to lower water requirements and precision farming using satellite imagery and global

[38] Inteview with Mr Polad-Polad Zade, First Deputy Minister of Minvodstroy (Ministry of Water Management Construction), Moscow, 14 September 1989.

[39] World Bank and the ICWC, *Developing a Regional Water Management Strategy: Issues and Work Plan*, Aral Sea Basin Program Technical Paper Series, Washington, DC, April 1996, pp. 24–5.

positioning system technology.[40] Israel, with similar climatic conditions to the Aral Sea basin, but technologically sophisticated irrigation practices, has an average withdrawal of 5,590 m^3/ha.[41] Such a figure is probably out of reach in the Aral Sea basin because of the much greater length of water-losing delivery canals and the different crop mix. However, if average water withdrawals could be lowered to 8,000 m^3/ha, as Victor Dukhovnyy, a leading Central Asian irrigation expert, proposed in 1985, gross savings of 36.5 km^3 and net savings of 27.7 km^3 (27 per cent of withdrawals for 1995) would accrue.[42]

Savings from efficiency improvements could well be significantly larger. Evidence suggests that the official figures provided by the basin states for their irrigation withdrawals are underestimates.[43] The system of measuring deliveries to cooperative users (the former collective and state farms), inadequate during Soviet times, has significantly deteriorated. Frequently, what is reported as farm usage is not based on actual measurement (because measuring equipment is absent or not working) but is an educated guess derived from the established water use norms for the region and farm crop mix (it represents what should be delivered to the farm, not what is actually used).

Usage by the variety of private farming types that have developed in the Aral Sea basin countries since independence is even more of a mystery. In the majority of instances, no organized system exists to measure their withdrawals. It is also likely that the area irrigated in any given year is exaggerated. It represents the area with completed irrigation facilities but is not adequately reduced for those systems that are under repair or not working or that have been removed from production (for example, because of excessive soil salinization or lack of water). Farms and the agricultural/water management hierarchy have an incentive to over-report the area irrigated and under-report water withdrawals. This makes them look better in terms of efficiency – that is, in water use per hectare.

Implementation of a large-scale programme for technical improvement of irrigation in the Aral Sea basin would be a gigantic undertaking and require a long period of implementation. For the basin as a whole in 1994, the length of main and inter-farm irrigation channels was around 48,000 km; only 28 per cent of these were lined to

[40] Some have proposed replacing part of the furrow irrigation systems, covering 70 per cent of the irrigated area, with drip irrigation which uses much less water. However, drip systems are very expensive, require a high level of maintenance and would be subject to severe plugging problems in the Aral Sea basin owing to the high sediment content of water in the Amu Dar'ya and Syr Dar'ya (conversation with Mr N. Jones, Meredith Jones Group, London, 25 January 2000).

[41] ICAS, *Fundamental Provisions*, Table 4.1.

[42] V. A. Dukhovnyy, 'Ekonomit' orositel'nuyu vodu!' [Save irrigation water!], *Gidrotekhnika i melioratsiya* [Hydrotechnology and Reclamation], No. 5, 1985, pp. 40–43.

[43] Conversations and discussions with Mr Onno Schapp, on-farm irrigation management specialist, WARMAP project of the European Union's TACIS programme, during the period October 1996 to August 1997 when the author was working as director of a USAID project in Tashkent, Uzbekistan.

reduce filtration.[44] The situation was even worse for on-farm canals: over 268,000 km with 21 per cent lined. In Uzbekistan 10,000 km of main and inter-farm canals need lining, and nearly 2 million ha with older irrigation systems (nearly half of the irrigated area) need reconstruction.[45]

Fifty per cent of irrigated lands in the basin suffer from salinity, which is probably the most serious problem faced by irrigated agriculture here.[46] In a 1989 survey of salinized lands, 53 per cent were considered slightly salinized, 32 per cent moderately salinized and 13 per cent strongly salinized. Large areas of irrigated lands also suffer from high water tables as a result of the lack of drainage facilities or facilities that are inadequate or not working properly. Drainage systems that keep the water table sufficiently deep are exceptionally important for irrigation in dry areas with saline ground water such as are common in the Aral Sea basin. Field experiments show, for example, that in Uzbekistan's Bukhara *oblast'* it took 26,000 m^3/ha to grow cotton and prevent secondary soil salinization with a depth to groundwater of one metre, but only 8,000 m^3/ha when the water table stood at 3 metres.[47] In the former case, most of the water was used to flush salts from the upper layers of the soil rather than to irrigate the crop.

Of the major irrigation zones, only the Golodnaya steppe and Khorezm oasis in Uzbekistan have adequate drainage facilities. Groundwater levels are rising nearly everywhere, exacerbating waterlogging and soil salinization. Disposal of the huge volumes of irrigation drainage water that is salinized and polluted with agricultural chemicals and fertilizers is also an increasingly serious problem. These discharges had reached 43 km^3 annually by the mid-1990s.[48]

Uzbekistani experts have estimated the rehabilitation costs for the older irrigation systems in the basin. Substantially improving system efficiency could cost $3–4,000/ha.[49] A World Bank-sponsored study indicates renovation of irrigation and drainage systems could run to $3000/ha.[50] This document also estimates that 5.4 million ha (68 per cent) of the 7.94 million ha of irrigated lands in 1995 need reconstruction. At $3,000/ha, this would cost $16 billion. Uzbekistan and Turkmenistan, with the largest areas under irrigation and the largest share of the systems needing reconstruction,

[44] World Bank, et al., *Developing a Regional Water Management Strategy*, p. 23.

[45] Antonov, 'O programme dal'neyshego', p. 9.

[46] Pankova et al., *Prirodnoye i antropogennoye*, pp. 85–7.

[47] V.I. Bobchenko, 'Obespechit' ekologicheskuyu nadezhnost' meliorativnykh system v oroshayemoy zone' [To ensure the ecological reliability of reclamation systems in the irrigated zone]', *Melioratsiya i vodnoye khozyaystvo* [Reclamation and Water Management], No. 6, 1989, p. 17.

[48] ICAS, *Fundamental Provisions*, Chapter 6.

[49] Report given by Mr G.N. Djalalov, Deputy Minister of Water Management for Uzbekistan, at the Working Meeting of Representatives of the Water Energy Services of Kazakhstan, Uzbekistan, Tajikistan and Kyrgyzstan for the preparation of recommendations for the efficient use of the resources of the Naryn-Syr Dar'ya cascade of reservoirs over the long term, Almaty, Kazakhstan, 1–2 October 1996.

[50] World Bank, et al., *Developing a Regional Water Management Strategy*, p. 25.

would bear the brunt of the bill. It is extremely doubtful whether the states of the basin, individually or collectively, have the funds now, or will in the near-term or medium-term future, to support so costly a project. Furthermore, the condition of irrigation systems in the basin has deteriorated since independence as funds for maintenance and repair have plummeted; responsibility for system maintenance has fallen between the cracks or been dumped on farm units which are unable or unwilling to conduct the necessary work; and the supply of replacement parts and equipment that formerly came from other republics of the USSR has dried up. Overall rehabilitation costs are steadily rising with time.

One disturbing result of the deteriorating condition of irrigated lands in the Aral Sea basin (plus other factors such as poor seed quality, a sharp drop in the use of fertilizers and pesticides, poorly timed agricultural operations and harvests, and lack of proper crop rotation) has been a steady decline in yields of most major irrigated crops, especially in Kazakhstan and Tajikistan.[51] Between 1990 and 1994 yields of cereals fell by 19 per cent in Uzbekistan, 37 per cent in Kazakhstan, 23 per cent in Turkmenistan, 50 per cent in Kyrgyzstan and 59 per cent in Tajikistan. Cotton yields declined by 7 per cent in Uzbekistan, 31 per cent in Kazakhstan, 2 per cent in Turkmenistan, 24 per cent in Kyrgyzstan and 31 per cent in Tajikistan. Vegetable yields rose by 23 per cent in Turkmenistan, held steady in Uzbekistan and fell by between 33 per cent and 68 per cent in the other states.

Switching the crop mix from high water use crops (rice and cotton) towards lower ones (grains, vegetables, melons, fruits and soya beans) could be a relatively inexpensive means of reducing water use compared with massive technical rehabilitation. In fact, crop substitution is the main reason why irrigation withdrawals per hectare substantially dropped between 1990 and 1995 (see Table 3). The main switch was from cotton to grains (mainly winter wheat). From 1990 to 1994, the percentage of the total cropped area devoted to grains increased from 12 per cent to 26 per cent, while cotton shrank from 40 per cent (or perhaps 45 per cent) to 32 per cent. The purpose of these changes, however, was not so much water savings but strengthening the national food bases. Uzbekistan made the largest absolute substitution of grain for cotton and Turkmenistan the largest percentage switch. It is unlikely that grains in these two states will replace further large areas of cotton in the near future, as cotton is an important foreign currency-earning export crop for both.

Adoption of government policies promoting irrigation water pricing and privatization of land, and giving rights of self-governance and responsibility for management of on-farm and inter-farm irrigation systems to farmer-irrigators, could encourage the introduction of water-saving practices in the Aral Sea basin without massive governmental

[51] ICAS, *Fundamental Provisions*, Chapter 10, Table 10.2.

expenditures.[52] Kazakhstan, Tajikistan and particularly Kyrgyzstan have taken some serious steps in this direction. Uzbekistan has talked about these subjects but made only feeble moves towards implementing meaningful policies. Turkmenistan has done practically nothing. Among the key obstacles to further progress along these lines are governmental resistance based on fear of losing social and economic control; opposition from the former collective (now cooperative) farms and local officials; fear of land speculation and exacerbating rural underemployment and unemployment; lack of means to measure water deliveries to farmers; and the impoverished state of the farming economy. These matters will be discussed in more detail in Chapter 6.

The major burden for reducing irrigation usage must rest on Uzbekistan as it has the majority of the irrigated area and irrigation withdrawals in the Aral Sea basin. Turkmenistan, which accounts for a significant share of the irrigated area and withdrawals in the Amu Dar'ya basin, could also make substantial contributions to water savings. Since Tajikistan, Kazakhstan and Kyrgyzstan irrigate much smaller portions of the basin and withdraw considerably less water, their potential contributions, although not insignificant, would be much smaller.

However, Tajikistan and Kazakhstan (see Table 5) have specific water use rates in irrigation that are higher than those of the other states, so they could make a disproportionately large addition to savings – the former by technical improvements and the latter by significantly reducing production of rice (with its very heavy water consumption) along the lower reaches and in the delta of the Syr Dar'ya. Afghanistan and Iran withdraw very little from the basin; their possible contribution to water savings is nil. This may change in the future for Afghanistan, however, as it could substantially increase its withdrawals from the Amu Dar'ya.

Improving irrigation to make more efficient use of basin water resources is not the only major problem that confronts the Central Asian states. Other issues are equally serious: equitable sharing of water resources between the upper riparians (Kyrgyzstan, Tajikistan and Afghanistan) and the lower riparians (Kazakhstan, Uzbekistan and Turkmenistan); contention over how the main upstream dams on the Amu Dar'ya and Syr Dar'ya should be operated (hydropower production or irrigation); and the interstate response to the Aral Sea crisis. These matters will be discussed in the next chapter.

[52] Information acquired by the author during a one-year assignment as Resident Advisor on Water and Environmental Management Policy to the Government of Uzbekistan, under USAID's Environmental Policy and Technology Project, September 1996 to October 1997; Philip P. Micklin, 'Development of Self-governing Irrigation Systems in Uzbekistan: Problems and Prospects', draft final report on the training seminar held in Tashkent, Uzbekistan, 29–30 April 1997, prepared for the Central Asia Mission US Agency for International Development, Almaty, Kazakhstan under Contract No. CCN-0003-Q-14-3165-00 of the Environmental Policy and Technology Project, 25 May 1997; Philip P. Micklin, 'Developing Water Pricing Systems for Uzbekistan: Key Policy Issues and Initial Steps', draft final report on the training seminar held in Khodjikent, Uzbekistan, 28 July–1 August 1997, prepared for the Central Asia Mission US Agency for International Development, Almaty, Kazakhstan under Contract No. CCN-0003-Q-14-3165-00 of the Environmental Policy and Technology Project, 11 August 1997.

5 SHARING THE WATERS OF THE ARAL SEA BASIN

Chapter 2 pointed out that the lion's share of the water resources of the Aral Sea basin is of an interstate (also termed transnational) character. These resources consist of rivers, lakes, usable groundwater and return flows from uses (predominantly irrigation) that cross or form national borders or directly affect water resources in other basin states. The most important specific resources are the Amu Dar'ya and Syr Dar'ya rivers, their major tributaries and the Aral Sea. How these entities are shared and/or managed already plays a crucial role in the political, economic and environmental future of the individual states and the region as a whole (and will continue to do so).

5.1 Institutional structures

The former Aral Sea basin republics of the Soviet Union realized as independence approached in late 1991 that they needed to establish institutional mechanisms to enhance cooperation in the management of interstate water resources.[1] The first major step was the signing in February 1992 of an agreement on the joint management and protection of interstate water resources. The agreement created an Interstate Commission on Water Coordination (ICWC) [*Mezhgosudarstvennaya koordinatsionnaya vodokhozyaystvennaya komissiya* in Russian] for overseeing the regulation, efficient use and protection of interstate watercourses and bodies.

The commission is composed of the heads of the main water management organizations in each republic or their designates (the Water Resources Committee in Kazakhstan, the Ministry of Agriculture and Water Management in Kyrgyzstan and Uzbekistan, the Ministry of Water Resources in Turkmenistan and Tajikistan). It meets several times each year to discuss and decide interstate water management policy issues. It also sets (and if necessary later adjusts) the allocation of water among the republics and to the Aral Sea and its deltas during the hydrologic year (October to October) on the basis of forecasts of water availability made by the hydrological and meteorological agencies in each country. The allocation scheme was a continuation of the system codified in 1984 and 1987 under the Ministry of Water Management of

[1] ICWC, *Central Asia 1992–1997*, Tashkent, 1997, published by the SIC of ICWC, pp. 4–8.

Table 6: Water withdrawal allocations for the 1996–7 hydrologic year made by the ICWC (km^3)

Country	Syr Dar'ya basin	%	Amu Dar'ya basin	%	Aral Sea basin	%
Tajikistan	1.65	7.4	7.90	12.9	9.55	11.4
Kyrgyzstan	0.18	0.8	0.15	0.2	0.33	0.4
Turkmenistan	0.00	0.0	22.00	35.9	22.00	26.3
Uzbekistan	9.46	42.1	22.00	35.9	31.46	37.5
Kazakhstan	6.72	29.9	0.00	0.0	6.72	8.0
Aral Sea region	4.43	19.7	9.30	15.2	13.73	16.4
Total	22.43	100.0	61.35	100.0	83.78	100.0

Source: Compiled from data in Nauhno-informatsionnyy tsentr Mezhgosudarstvennoy koordinatsionnoy vodokhozyaystvennoy komissii (MKVK), *Tsentralnoy Azii* [ICWC of Central Asia], Bulletin 14, p. 22.

the former USSR. Responsibility for determining the operating regimes for the reservoirs along the interstate rivers was also placed on the ICWC. Inter-republic water management disputes are to be decided by the commission, with help from a neutral arbitrator, if needed. The ICWC consists of a secretariat (in Khodjent, Tajikistan), scientific information centre (in Tashkent, Uzbekistan) and the basin management authorities (BVOs) [*Basseynovoye vodokhozyaystvennoye ob'yedinennoye*] for the Syr Dar'ya (in Tashkent) and Amu Dar'ya (in Urgench, Uzbekistan). The BVOs were first established by the USSR in 1986.[2] Since independence, they have been charged with managing and monitoring the allocations made by the ICWC to member states.

The ICWC makes annual withdrawal allocations based on estimates of water availability for the ensuing hydrologic year. Corrections may be made at the regular quarterly meetings or at specially convened sessions. As was the case under the Soviet system, the water-sharing scheme is heavily tilted towards irrigation and the interests of the downstream riparian states. Table 6 shows allocations for the 1996–7 hydrologic year. The two leading irrigators, Uzbekistan and Turkmenistan, were assigned 38 per cent and 26 per cent, respectively, of the total, with each to receive 36 per cent of the flow of the Amu Dar'ya, and Uzbekistan to get an additional 42 per cent of the Syr Dar'ya's discharge. The ICWC allocated Kazakhstan 30 per cent of withdrawals from the Syr Dar'ya.

The upstream flow generating states were given the residue. Tajikistan was allocated 11 per cent: 7 per cent of the withdrawals from the Syr Dar'ya and 13 per

[2] Mezhdunarodnyy Fond spaseniya Arala, *Programma basseyna Aral'skogo Morya: proshloye i vzglada v budushcheye* [Aral Sea Basin Program: the past and a look toward the future], faza 1, otchet No. 4 (o realizatsii rabot) [Phase 1, Report No. 4 (on the realization of works)], Tashkent, June 1997, p. 7.

cent from the Amu Dar'ya. Kyrgyzstan was given only 0.4 per cent, representing less than 1 per cent of the withdrawals from the Syr Dar'ya and 0.2 per cent from the Amu Dar'ya. The ICWC allotted 16 per cent of the withdrawals (in this case, flow) of the Syr Dar'ya and Amu Dar'ya to the near Aral Sea region. In Soviet times, the near Aral Sea region did not have a specific allocation. The allotted water is intended for both the deltas and the sea and much less water than might appear reaches the sea proper owing to major diversions into the delta for ecological purposes (e.g. support of deltaic lakes and reservoirs), water supply and even irrigation.[3]

The low allocation for Kyrgyzstan from the Syr Dar'ya is puzzling. In an average flow year, it is supposed to receive 4.90 km^3 or about 13 per cent.[4] One would have expected Kyrgyzstan to receive around 3 km^3 in the 1996–7 hydrologic year (a below average flow year for the Syr Dar'ya). An explanation may be that Kyrgyzstan withdraws some water from this river. The Syr Dar'ya is considered a national resource and is not included in ICWC allocations as the commission is only empowered to allocate river flow that is considered interstate in character.

There are two caveats about the allocation data. First, the European Union's Water Resources Management and Agricultural Production in the Central Asian Republics (WARMAP) project analysis for the period 1994/5–1996/7 indicates missing data and inconsistencies with two other comparable data sets (those developed by the BVOs and the database developed independently by the EU's Water Resources Management Information System project).[5] Part of the inconsistencies may result from changes in ICWC reporting practices from year to year. Second, the actual total withdrawals and withdrawals by each state may differ somewhat from the allocations in any given year; 1995–6 and probably 1996–7 were dry and much less than the allocated share of water reached the Aral Sea region as upstream irrigation took more than its share.

As might be expected, the upstream states, particularly Kyrgyzstan, have complained about the allocation scheme. Their mountain territories generate 80 per cent of the flow in the basin of the Aral Sea, yet their allowed withdrawals are small compared with the share assigned to Uzbekistan, Kazakhstan and Turkmenistan. Furthermore, to serve irrigation, most of the flow must be accumulated in reservoirs on their territories and released during the growing season, restricting the ability of Kyrgyzstan and Tajikistan to generate winter hydroelectricity. This is a particularly serious problem for Kyrgyzstan since it has faced chronic winter energy shortages in the 1990s, as deliveries of coal (from Kazakhstan) and gas (from Uzbekistan) have, at

[3] WARMAP Project, *Formulation and Analysis of Regional Strategies on Land & Water Management*, July 1997, p. 72.

[4] T. Sarbayev, 'O dal'neyshem razvitii orosheniya v Kyrgyzstane' [On the further development of irrigation in Kyrgyzstan], *Vestnik Arala*, No. 1 (vodnyye resursy), Spring 1996, p. 11.

[5] WARMAP project, *Formulation and Analysis*, pp. 72–84.

best, been irregular. During the Soviet era such deliveries were guaranteed to cover Kyrgyzstan's opportunity costs in forgoing winter hydroelectric generation for the benefit of the downstream irrigating republics.

The 1992 agreement has become acutely contentious among the five states. Turkmenistan, Kazakhstan and especially Uzbekistan want to see the existing allocation schemes and operational regimes continued.[6] Kyrgyzstan and Tajikistan want considerably more of the flow allocated to them (so they can expand irrigation) and more freedom to generate winter hydropower. Over the past five years Kyrgyzstan has repeatedly violated winter release limits (set by the ICWC) at the huge Toktogul reservoir on the Naryn, the chief tributary of the Syr Dar'ya. This has reduced water available during the following summer season for downstream irrigation. It has also caused winter flooding in Uzbekistan and Kazakhstan, forcing emergency diversion of flow, which should go from the Aral Sea to Lake Arnasay in the desert. Charges have been levelled that Kyrgyzstan is not releasing the right quantity of irrigation water from Toktogul in the summer.

There are also allocation conflicts among the downstream states. Most contentious is the disagreement between Uzbekistan and Turkmenistan over the Kara-Kum canal. Under construction since the mid-1950s, the almost 1400-km long canal is allocated 13 km^3 annually from the Amu Dar'ya.[7] This facility irrigates almost 1 million ha in Turkmenistan and is the source of municipal water supply for the capital, Ashgabat. The Turkmen government considers the canal fundamental to national survival and is intent on lengthening it and irrigating even larger areas. On the other hand, the Uzbek government views the unlined Kara-Kum as a man-made river flowing through the desert that loses huge amounts of water to filtration. It also sees the canal as one of the key factors contributing to the demise of the Aral Sea.[8] The Uzbeks are adamantly opposed to further construction on the canal and increased water diversions to it. Owing to lack of maintenance, the canal is rapidly deteriorating, further increasing water losses.

In March 1993, the presidents of the five republics signed an agreement to promote cooperation in solving the key problems related to the Aral Sea and surrounding region.[9] This established the Interstate Council on the Problems of the Aral Sea Basin

[6] Information acquired by the author during a one-year assignment as Resident Advisor on Water and Environmental Management Policy to the Government of Uzbekistan, under USAID's Environmental Policy and Technology Project, September 1996 to October 1997.

[7] Tim Hannan and Sarah L. O'Hara, 'Managing Turkmenistan's Kara Kum Canal: Problems and Prospects', *Post-Soviet Geography and Economics*, Vol. 39, No. 4 (April 1998), pp. 225–35.

[8] Philip P. Micklin, *The Water Management Crisis in Soviet Central Asia*, The Carl Beck Papers in Russian and East European Studies, No. 905 (Pittsburgh, PA: The Center for Russian and East European Studies, August 1991), p. 46.

[9] Agreement on joint activities in addressing the Aral Sea and the zone around the sea crisis, improving the environment and ensuring the social and economic development of the Aral Sea region, signed in Kzyl-Orda, Republic of Kazakhstan, 26 March 1993.

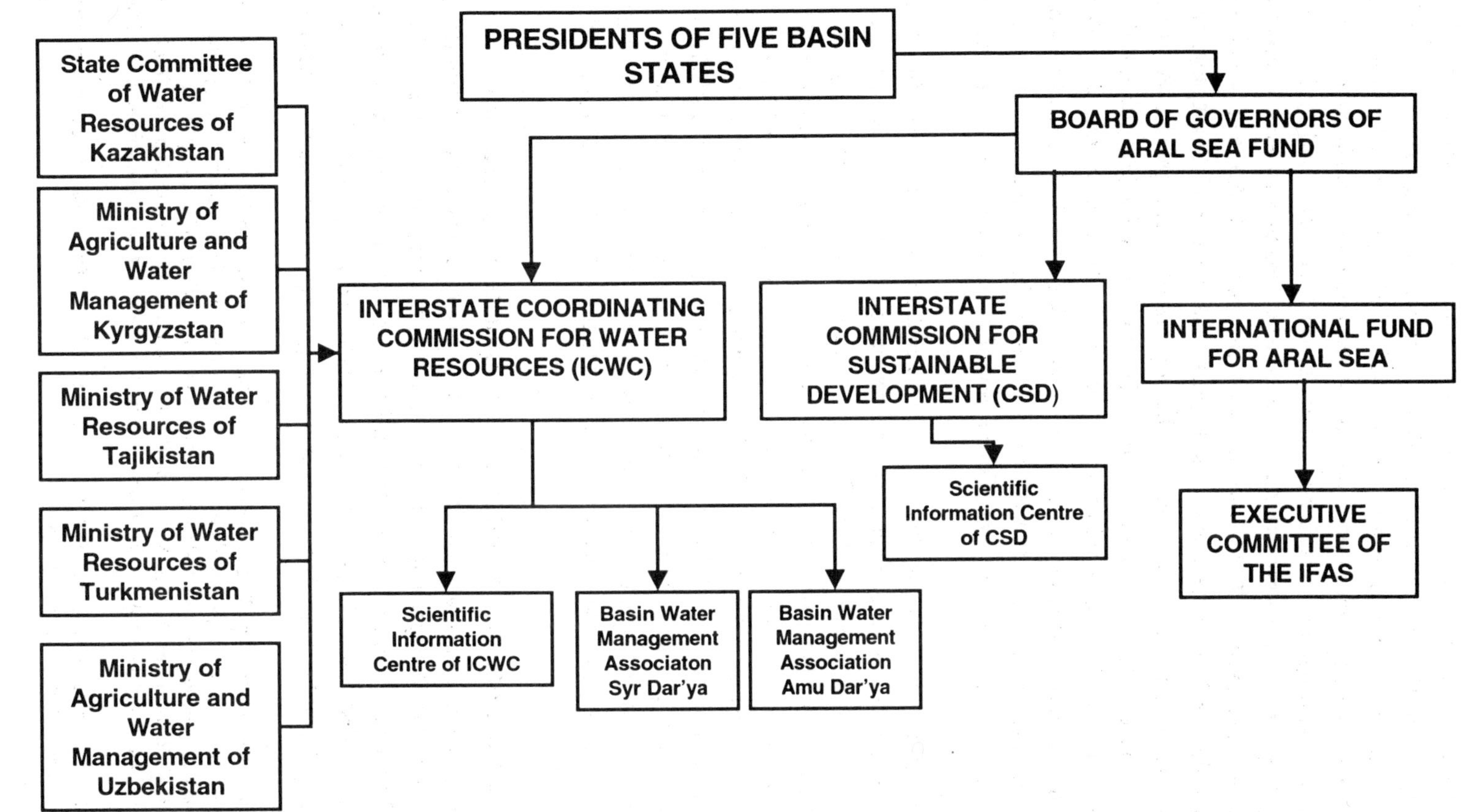

Figure 3: The structure of interstate organizations for addressing the Aral Sea crisis

(ICAS) [*Mezhgosudarstvennyy Sovet po problemam basseyna Aral'skogo morya*]. A major purpose of the new organization was to facilitate assistance from the World Bank and other international donors.[10] Composed of 25 members, five from each state, ICAS assumed management of various Aral Sea basin assistance programmes. It was to meet at irregular intervals (six meetings were held between July 1993 and February 1997), but day-to-day activities were to be handled by an executive council in Tashkent.[11] The existing ICWC was made a component of ICAS, and in January 1994 a sustainable development commission (SDC), headquartered in Ashgabat, was added to the administrative structure of the organization. The presidents of the republics also created an international fund for the Aral Sea (IFAS) [*Mezhdunarodnyy Fond spaseniya Arala*] composed of ten members, two from each state, to collect funds from each basin state (the recommended contribution was 1 per cent of national income) for financing rehabilitation efforts. The fund's headquarters were situated in Almaty, Kazakhstan, and President Nazarbayev of Kazakhstan was elected as the first head.

The presidents made major changes in ICAS and IFAS in February 1997. The most important step was abolishing ICAS and merging the functions of the two organizations into a restructured IFAS to reduce duplication of effort, simplify the administrative structure, overcome bureaucratic inertia and revitalize improvement efforts in the Aral Sea basin. The new structure of interstate organizations for dealing with the Aral crisis is shown in Figure 3. The leadership of IFAS is to rotate in a two-year cycle among the Central Asian heads of state, beginning with Islam Karimov, president of Uzbekistan.[12] Annual financial contributions to the fund were also revised downward to 0.3 per cent for richer downstream Uzbekistan, Kazakhstan and Turkmenistan, and to 0.1 per cent for poorer upstream Kyrgyzstan and Tajikistan.

5.2 International donor efforts to promote cooperation

Since independence, international aid donors have played a major role in promoting cooperation in the management of the transnational water resources in the Aral Sea basin.[13] The World Bank (the International Bank for Reconstruction and Development) was the first major player on the scene. In 1992 and 1993, the Bank formulated an Aral Sea Basin Program (ASBP) which was to be carried out in three phases over

[10] P. Micklin, 'International and Regional Responses to the Aral Crisis: An Overview of Efforts and Accomplishments', *Post-Soviet Geography and Economics*, Vol. 39, No. 7 (September 1998), pp. 406–9.

[11] Mezhdunarodnyy Fond spaseniya Arala, *Programma basseyna Aral'skogo Morya*, pp. 4–11.

[12] Ibid., pp. 41–2; World Bank, *Aral Sea Basin Program (Kazakhstan, Kyrgyz Republic, Tajikistan, Turkmenistan and Uzbekistan): Water and Environmental Management Project* (Washington, DC, May 1998), p. 9.

[13] See Micklin, 'International and Regional Responses', pp. 399–417, for a detailed treatment of international assistance efforts in the Aral Sea basin.

15–20 years and expected to cost around $250 million (as of 1996, the estimated cost had risen to $470 million).[14] The main goals of the programme are:

(1) the rehabilitation and development of the Aral Sea disaster zone;
(2) strategic planning and comprehensive management of the water resources of the Amu Dar'ya and Syr Dar'ya;
(3) building institutions for planning and implementing the above programmes.

The Bank encouraged the basin states to create ICAS and IFAS and has worked with and through these organizations to realize the ASBP. Afghanistan was invited to join the ASBP but did not respond.[15]

The preparatory portion of the first phase was supposed to take 18 months and be finished by the end of 1995.[16] The Bank declared these activities finished in mid-1997, more than a year behind schedule, even though parts of the planning process were still under way. The chief reasons for the delay appear to be lack of experience working in what was, for the Bank, the new social, political and economic environment of Central Asia, and laggard funding of the preparatory work (as of mid-1997 only $15.4 million of project financing was completed or under way).

The Bank undertook a review of the ASBP in July 1996 in which donor and regional representatives participated.[17] While recognizing the accomplishments and progress made, the review proposed a number of major changes. The basin states were advised to strengthen the role of regional institutions and depend less on donor influence; to increase their political and financial commitments to the regional bodies; to clarify the priorities between national and regional tasks; to focus more attention on activities which could be quickly implemented; and to make clearer the roles of the different regional entities (ICAS, IFAS, ICWC, SDC and BVOs). The basin states recommended that the Bank concentrate more on financing projects supportive of the ASBP, gradually reduce its technical assistance role and speedily prepare the follow-on project to the first phase of the ASBP.

The next phase in the Bank's Aral Sea Basin Program is the Water and Environmental Management Project.[18] Initiated in August 1998, it is slated to run for four years at a cost of $21.2 million. In line with the new emphasis on regional responsibility for the ASBP, the executive committee of IFAS is managing the programme, with the Bank playing a cooperative/advisory role. Most of the components of this plan are directly intended to improve the management of transnational water resources. The

[14] Ibid., pp. 406–9.
[15] World Bank, *Aral Sea Basin Program*, p. 9, footnote 16.
[16] Ibid., p. 9.
[17] Ibid., pp. 8–11.
[18] Ibid., pp. 19–34.

most important of these are the formulation of national and regional water and salt management strategies and agreements, and the development of low-cost, local, on-farm water conservation measures. The purpose is to improve the efficiency of water use, deal with the increasingly serious problem of salinization of soils from improper irrigation practices and reduce the amount of irrigation drainage water flowing back into rivers.

A second programme aims at the strengthening of existing interstate water-sharing agreements. A third component entails a public awareness campaign to sensitize the public, water suppliers and water users to the key issues and strategies of the ASBP for saving water and to effect water-saving behavioural change. Dam and reservoir management constitutes a fourth part of the project. This is intended to ensure the sustainability and safety of dam and reservoir infrastructure along the interstate rivers. A fifth undertaking is the transboundary monitoring of water quality and quantity, which is a precondition of more effective and accepted interstate water management/sharing agreements. This is of great importance as the network of hydrologic and water quality observation stations established under the USSR is rapidly deteriorating.[19]

Besides those of the Bank, there are several other international efforts to improve management of interstate water resources in the Aral Sea basin. The United States, through the United States Agency for International Development (USAID), funded the Environmental Policy and Technology (EPT) project. This ran from 1993 to 1998 and supported regional efforts to reach agreement on operation of the Toktogul dam and reservoir on the Syr Dar'ya.[20] Located in Kyrgyzstan, this huge storage facility controls downstream water use in Uzbekistan and Kazakhstan. Controversy over the appropriate operating regime (hydropower, irrigation or something between the two) between upstream Kyrgyzstan and downstream Uzbekistan and Kazakhstan had reached crisis point. Working with the Interstate Council for Kazakhstan, Kyrgyzstan and Uzbekistan (ICKKU), the EPT project helped the three countries reach a framework agreement that was signed in March 1998.[21] A smaller-scale follow-up project sponsored by USAID is under way to help the ICKKU implement the agreement.[22]

The UN has been providing a variety of assistance to the Aral Sea basin since 1990.[23] The most ambitious UN efforts in recent years are proceeding under the direction of the UN Development Programme (UNDP). One of UNDP's major efforts is aimed at strengthening the regional organizations that have been established

[19] World Bank et al., p. 15.

[20] Micklin, 'International and Regional Responses', pp. 409–10.

[21] World Bank, *Aral Sea Basin Program,* p. 21; Environmental Policies and Institutions for Central Asia (EPIC) programme, sponsored by USAID, 'Information Summary', 13 September 1999, 7 pages.

[22] Ibid.

[23] Micklin, 'International and Regional Responses', pp. 410–11.

to deal with the Aral crisis (earlier versions of ICAS and IFAS, and now the reconstituted IFAS).

The EU initiated WARMAP, a major aid programme for the Aral Sea basin states, in 1995.[24] The objectives of the programme were

> to provide the administrative and technical framework within which the five republics of the Aral Sea basin can develop policies, strategies and development programmes for utilization, allocation and management of the water resources of the basin; and to assist at the regional level with the establishment of the institutional structure required to prepare and execute policies and strategies to give effect to the agreed framework on water allocation and management.[25]

Phases one and two were completed by mid-1997 at a cost of around $6 million.

A number of activities directly related to management of interstate water resources were completed over this period. These included the development of a geographic information system land and water database for the basin, providing help to the World Bank and IFAS in their efforts to improve and legally codify the 1992 interstate water-sharing agreement; the sponsorship and funding of training seminars and workshops; and an attempt to gather detailed data on irrigated water use at the farm level through a water use and farm management monitoring survey.[26] A continuation/follow-on project to WARMAP began in early 1998 and will run for several more years.[27]

5.3 Improvements needed

The five former USSR Central Asian states, to their credit, have created regional institutions (IFAS, ICWC and ICKKU) to deal with questions of basin-wide and transboundary concern (including allocation of water from the international rivers, Amu Dar'ya and Syr Dar'ya), and the human and ecological problems connected with the desiccation of the Aral Sea. So far, the Central Asian states have approached these matters in a spirit of mutual cooperation and respect for national rights. Furthermore, although these new institutions, so far, have not been able to find a satisfactory resolution of the most critical interstate water management conflicts, they serve as a forum for reasoned discussion of these matters and act as a safety valve to defuse tensions over water issues among the states.

[24] Ibid., p. 411.

[25] European Commission, *Water Resources Management and Agricultural Production in the Central Asian Republics: WARMAP PROJECT*, Phase 1: Project Preparation Reports, Executive Summary (Tashkent, September 1995), p. 1.

[26] World Bank, *Aral Sea Basin Program*, pp. 8–9.

[27] Information acquired by the author during a two-month assignment as consultant to the National Sustainable Development Commission of Uzbekistan, May to July 1998.

Since the break-up of the Soviet Union, the international community has become deeply involved in assisting the basin states in their efforts to cooperate on water sharing and to overcome the most serious problems associated with the Aral Sea. Donor aid has focused on several key issues:

(1) improving the human conditions and ecological situation in the zone around the sea that has been most seriously afflicted;
(2) raising the efficiency of water use in irrigated agriculture in the basin;
(3) building the capacity of the regional institutions created to deal with Aral Sea basin issues to make them more effective and efficient in carrying out their duties;
(4) promoting the development of agreements on the sharing of the waters of international rivers and operation of the Toktogul reservoir that comport with international legal standards and norms.

In spite of the successes enumerated above, many critical issues await resolution. The regional organizations (IFAS, ICWC and ICKKU) need to have their roles clarified. The responsibilities and functions of the three overlap in key areas of management and programme implementation. This has led to confusion, disagreements and costly duplication of effort that has slowed the introduction of improvement measures. A sincere effort should be made to include Afghanistan in negotiations about water allocations from the Amu Dar'ya, as the country accounts for about 8 per cent of the river's flow.

IFAS also needs to receive more support from the member countries in terms of financing, recognition of its interstate status and the supply of qualified personnel. Lack of adequate funding has been a serious hindrance to the work of this organization. On the other hand, the Aral Sea basin states have invested in water management and Aral Sea zone improvement efforts on a national basis. The executive committee of IFAS reports that Uzbekistan, Kazakhstan and Turkmenistan have together been spending around $650 million annually on socio-economic and environmental stabilization efforts in the disaster zone around the sea.[28] This figure is substantial, although probably exaggerated, particularly by Uzbekistan's use of an artificially high exchange rate between the national currency (*sum*) and the US dollar.

Several changes in the operations of the international donors need implementation. Of primary importance is improved cooperation among the major players such as the World Bank, UN, United States and EU. Talk abounds of 'working together' to minimize programme duplication and costs as well as maximizing results. However, this author's experience of working in the donor community in Tashkent during 1996–7 and summer 1998 suggests such rhetoric signifies good intentions more than

[28] World Bank, *Aral Sea Basin Program*, p. 12.

reality.[29] Given the large number of donors and the complexity of programmes under way in the Aral basin and the inevitable duplication of effort and conflicts that arise, it would be advisable to establish a high-level council of donors that would meet regularly (perhaps at least twice each year) to discuss the activities of the different donors and to facilitate field-level coordination and cooperation.

The laggard pace of donor programme implementation also needs attention. Expensive feasibility and planning studies drag on and, in the eyes of many Central Asians, there are no tangible results. Certainly, this is partially a result of the hindrances faced by the donor community in working within societies undergoing major economic, social and political transition. But a real effort is needed to move more quickly from the feasibility study and design phase to the completion of projects that make a difference in the lives of the region's people. The World Bank, as a result of its July 1996 review, has recognized this failing and is attempting to correct the problem in its new ASBP efforts.[30]

Another problem is that the donor community may, inadvertently, be developing an 'international welfare mentality' among the aid recipients in the Aral Sea basin. Frequently regional, national and local institutions in Central Asia approach international donors seeking funding for endeavours that are worthwhile but should really be funded from within the region. Often, it is not a case of lack of money but of governmental choices about spending priorities (e.g. the precedence of grandiose public buildings and monuments in the capital cities over new hospitals, clinics and drinking water supply facilities in the Aral Sea disaster zone). This is helped along by a pervasive belief that the Aral Sea situation is a world problem and that the international community has a moral and ethical obligation to help solve it.

Although Central Asia often complains that the international community is not doing enough, donors have provided substantial assistance. It is difficult to estimate the overall contributions because of the lack of comprehensive data and double counting (the World Bank commonly includes in its figures funding of Aral Sea basin efforts that are also counted by individual donors such as the United States). But in addition to other Aral Sea basin efforts, the Bank alone has, or is planning through fiscal year 2000, 16 loans and credits to Central Asian states of $605 million for improvement of land and water management. Other donors have given or plan to provide hundreds of millions more.

In the next chapter, we examine how the basin states formed from the USSR have coped with the policies and politics of managing the interrelated issues of agricultural land and water used for irrigation since independence. The success or failure of these efforts is as important as solving interstate issues in determining the region's future.

[29] Information acquired by the author during a one-year assignment and a two-month assignment, respectively.

[30] World Bank, *Aral Sea Basin Program*, Tables 4 ,5, 6.

6 NATIONAL LAND AND WATER POLICIES AND POLITICS

6.1 The Soviet legacy

The Aral Sea basin states formed from the Soviet Union at the end of 1991 (Kazakhstan, Uzbekistan, Kyrgystan, Tajikistan and Turkmenistan) faced a difficult future in many major regards (economic, political and cultural). Management of land and water was certainly a primary, if not the primary, issue among these. Since the late 1920s and early 1930s, the Soviet state had created institutions and an infrastructure of land and water management suited to national, not regional or local needs. The fundamental purpose was the steady, at times rapid, expansion of irrigation with a pre-eminent focus on the production of raw cotton. Nearly all this cotton was destined for export from the region to European Russia for manufacture into textiles and apparel or for export abroad.[1] In essence, the republics of the Aral Sea basin were transformed into raw-material appendages of European Russia.

Marxist–Leninist economics and ideology wedded to Stalinist collectivization and industrialization had destroyed traditional irrigated agriculture in the region. As discussed above (Chapter 5), traditional systems, although characterized by centralized, hierarchical control, devolved management responsibility, and the benefits from this, to the local level. Individual farming families using traditional technology carefully tended small, irrigated fields. They obtained surprisingly high yields with low water usage, as there were powerful incentives for this. A strictly enforced set of customs and rules ensured observance of land and water use rights and maintenance of irrigation systems.

Regional communist leaders, at the insistence of Moscow, replaced this system, evolved over centuries, with huge *kolkhozy* (collective farms) and *sovkhozy* (state farms) in which farmers were residual claimants to their own product (in the *kolkhozy*) or were paid wages (in the *sovkhozy* and, also, during the last few decades of the USSR, the *kolkhozy*). There were no incentives for farmers to work hard or take care of the land, as the land and its produce now belonged to the state. The ecologically sound system of small, irrigated fields was converted into huge tracts that were difficult to irrigate and subject to waterlogging and secondary soil

[1] R.D. Liebowitz, 'Soviet Geographical Imbalances and Soviet Central Asia', in R. A. Lewis (ed.), *Geographic Perspectives on Soviet Central Asia* (New York: Routledge, 1992), p. 120.

salinization. This, along with the provision of irrigation water without charge, led to a wasteful and rapid rise in water use per hectare. Giant water distribution systems were built rapidly, often shoddily, and managed by central authorities, with little sensitivity to local conditions. By the end of the Soviet period, water management was in crisis; the water resources of the Amu Dar'ya and Syr Dar'ya were essentially exhausted and the Aral Sea was rapidly drying, with a plethora of accompanying environmental and human problems.

6.2 Institutional and economic reforms

The most important agricultural land and water management issues faced by the new nations of the Aral Sea basin are discussed below. The primary emphasis is on Uzbekistan. It is the key nation in the basin, with 25 per cent of its area, 50 per cent of its population, 53 per cent of its irrigated area and 53 per cent of irrigation withdrawals (see Tables 1, 3 and 4). By the mid-1990s, irrigation in Uzbekistan, although accounting for only 15 per cent of agricultural land, provided 95 per cent of crop production.[2] Agriculture accounted for 30 per cent of GDP and over 40 per cent of employment.[3] Furthermore, during Soviet days, Uzbekistan was Moscow's favourite among the Central Asian republics and constituted the linchpin of irrigation development for the Aral Sea basin. How water for irrigation is managed here will have a powerful influence on its management throughout the region.

6.2.1 Land reform and farm restructuring

During the Soviet era, all land was owned by the state and the dominant forms of agricultural organization were the collective and state farm. From the beginning of the independence period, the governments of all the new nations understood the need to break these huge entities into smaller, more manageable units as well as to give farmers incentives to be more productive. The five nations have moved to dismantle the collective system of agriculture, but in different ways and at different paces.

Kazakhstan and Kyrgyzstan have come the farthest and fastest. Between 1991 and April 1997, Kazakhstan privatized 96 per cent of state agricultural entities, breaking up the state and collective farms into much smaller industrial cooperatives, economic

[2] Ministerstvo innostrannykh del, Gosudarstvennyy komitet respubliki Uzbekistan po okhrane prirody [Ministry of Foreign Affairs, State Committee of the Republic of Uzbekistan on the Environment], *Sostoyaniye okruzhayushchey prirodnoy sredy i ekologicheskiye problemy Respubliki Uzbekistan* [Condition of the natural environment and ecological problems of the Republic of Uzbekistan] (Tashkent, 1996), p. 13.

[3] US Department of Commerce, 'Uzbekistan: Economic Overview', April 1998, BISNIS Home Page (*http://www.bisnis.doc.gov/bisnis/country/984uzeco.htm*), p. 2; World Bank, 'Countries: Uzbekistan', August 1998 (*http://www.worldbank.org/jtml/estdr/offre/eca/uz2.htm*), p. 1.

associations and individually held farms.[4] Farmers were given property shares for the land that they used. Kyrgyzstan, over the same period, gave 90 per cent of agricultural land to farmers under a system of lifetime heritable tenure, with the right to sell the use of the land.[5] However, in neither country do farmers have the right to sell the land itself, a fundamental concept of property rights in the West. In spite of vocal opposition (substantial parts of the rural population continue to support the idea of retaining some form of collective agriculture and state ownership of land), both countries are moving towards legally establishing the right to buy and sell agricultural land and the formation of land markets for this.[6]

For much of the 1990s Tajikistan was embroiled in civil war, which hindered land reform. The government's goal is to provide land to individual farmers to reinvigorate agricultural production.[7] However, the agrarian sector is still dominated by large collective and state farms. These are to be transformed into cooperatives or associations in which individual farmers or families will have the right of inheritance leasing. The state retains a system of state orders for cotton and some other crops, but is gradually reducing the percentage that it takes and allowing farmers to sell the residue as they see fit.

Uzbekistan and Turkmenistan lag considerably behind Kyrgyzstan and Kazakhstan on the road to real land reform. All agricultural land continues to belong to the state and neither has announced firm plans to move towards private ownership with attached right of sale within the foreseeable future. Turkmenistan has made the least progress.[8] Collective and state farms still dominate its rural landscape, retarding incentives for farmers to be more productive. The state maintains the Soviet system of state orders (*goszakazy*) and controls the crop mix and pattern as well as purchase prices. Farmers are able to lease land from the state on an individual or family basis but few have done so.

Agricultural reform began in Uzbekistan in 1989, prior to independence, with a major expansion of the area of household plots.[9] These lands, averaging about 0.12 ha, were granted for use to peasant families. Farmers chose their own crops and sold their produce in the local markets, so the incentives for maximizing output were

[4] US Department of Commerce, 'Commercial Overview of Kazakhstan: Chapter 2 – Economic Overview and Outlook', June 1998, BISNIS Home Page. (*http://www.bisnis.doc.gov/bisnis/country/kzov982.htm*), p. 7.
[5] US Department of Commerce, 'Economic and Trade Overview of Kyrgyzstan: Part 3 – Economy', June 1998, BISNIS Home Page (*http://www.bisnis.doc.gov/bisnis/country/9806kyr3.htm*), p. 1.
[6] 'Kazakhstan President Puts off Sale of Farm Land', *Central Asia Online*, No. 41, 23 July–9 August 1999; 'Kyrgyz President Assesses Agricultural Sector', *Central Asia Online*, No. 28, 16 April 1999.
[7] US Department of Commerce, 'Commercial Overview of Tajikistan: Part 2 – Economic Overview', June 1998, BISNIS Home Page (*http://www.bisnis.doc.gov/bisnis/country/9806tjo2.htm*).
[8] US Department of Commerce, 'Commercial Overview of Turkmenistan: Part 2 – Economic Overview, Reforms', June 1998, BISNIS Home Page (*http://www.bisnis.doc.gov/bisnis/country/9806trk2.htm*), p. 3.
[9] Z. Lerman, J. Garcia-Garcia and Dennis Wichelins, 'Land and Water Policies in Uzbekistan', *Post-Soviet Geography and Economics*, Vol. 37, No. 3, March 1996, p. 147–67.

strong. During the late Soviet period, these farmers, although using only 3 per cent of arable lands, provided 20–25 per cent of gross agricultural production, primarily non-grain foodstuffs and meat. The government also granted more operational autonomy to state and collective farms as a means to stimulate greater production.

Between 1991 and 1996, some 25 laws (passed by the parliament, *Oliy Majlis* in Uzbek) and decrees (issued by the President Islam Karimov) dealt with key issues such as land use rights, property rights, status of state and collective farms, land leases by private farmers.[10] As a result, all state farms were transformed into collective farms, also known as cooperative farms, and leased farms. The former have permanent possession of lands they farm while the latter had ten-year minimum use rights. Both organizations pay a land tax or land rent to the state. The government also instituted a system for leasing land to individuals and families, which gave small-scale private farming a boost. By early 1997, private farms of all types and sizes numbered some 20,000 and occupied 16 per cent of the irrigated area, with more than 109,000 workers.[11]

In June 1998, the Uzbekistan government promulgated a new and very complex land code.[12] It retained the cooperative (collective) farm, renamed *shirkat*, as the basic unit for production of commercial crops such as cotton and grain. Members of the *shirkat* are given ownership shares in the operation. The government continued the system of leasing land for private agricultural use, with leases to run from 10 to 50 years. The new code also emphasized small-scale farming (*dekhan* or peasant farms) to be conducted by family members on lands allocated to them by the state. Although the new land code is a step in the right direction, it still prohibits private land ownership in the Western sense (the right to buy and sell) and continues the primary role of the government in the management of agriculture.

Despite reforms during the 1990s, collective and leased farms remain under state domination. They receive inputs from and sell a substantial part of their output to state-controlled companies. The Uzbek government retained the system of state orders for the two key crops, cotton and grain, following independence. However, these were reduced from 75 per cent to 30 per cent of assigned production targets between 1994 and 1997 and were supposed to be phased out completely in 1998.[13] As

[10] Food and Agricultural Policy Unit (FAPU) of the TACIS programme of the European Union, *Land Reform Policy in the Republic of Uzbekistan*, 11 June 1996, Tashkent, pp. 10–15.

[11] 'Concerning measures for state support of subsidiary small holding and peasant farms and increasing their role in food production to the country', Decree of the President of Uzbekistan, 18 March 1997 (translated from Russian by the TACIS/PIDEP project, 27 March 1997).

[12] 'Zemel'nyy kodeks Respubliki Uzbekistan' [Land code of the Republic of Uzbekistan], *Novoye slovo*, 2 June 1998, pp. 3–6.

[13] US Department of Commerce, 'Uzbekistan: Economic Overview', BISNIS electronic broadcast, 10 March 1997; Food and Agricultural Policy Unit (FAPU) of the TACIS programme of the European Union, *Marketing and Price Policy in the Republic of Uzbekistan*, 20 March, p. 3.

the state has a monopoly on cotton purchases, removal of state orders does not mean as much as it might seem, since cotton growers have little or no access to alternative buyers. The government set the purchase price of cotton subject to state orders at 60 per cent of the world level in 1995 and it was scheduled to reach parity with world prices in 1998.[14] After meeting state orders, cooperative and leased farms could sell their remaining produce to the state at negotiated prices. The state makes payments to cooperative and leased farms by transfer through the state-controlled banking system rather than in cash. This provides another lever of control over these organizations.[15]

6.2.2 Problems of private irrigated farming

Although the governments of the Aral basin states ostensibly encourage private irrigated farming as a key means to raise production and the efficiency of water use, it faces a myriad of problems.[16] Private farmers are generally impoverished. Without access to credit, they find it difficult to obtain basic production inputs (seed, fertilizer, equipment) or to maintain on-farm irrigation systems, let alone improve them. Although in theory farmers are eligible for bank loans, in practice these are complicated and difficult to obtain.

A second serious problem in Uzbekistan, Turkmenistan and Tajikistan, where the collective farm system has been retained to one degree or another, is the dependence of private irrigated farms on the collective and the *rayon* (district) political administration.[17] Private farmers (leaseholder or peasant) receive land from the collective farm with the approval of the local administration and, at least in Uzbekistan, the collective, can take it back from the farmer, again with the assent of the local political leader (the *hakim*). Reasons for cancelling rights to land use in Uzbekistan include breach of agreement, improper or irrational land use, and non-payment of land taxes. In Uzbekistan, collectives and local administrations have commonly charged private farmers with improper/irrational land use simply because the private farmers did not plant the crops that they were asked to plant.

The collectives also supply the private farmer with seed, equipment, technical advice and, perhaps most importantly, irrigation water which the farmer receives

[14] Economist Intelligence Unit, 'Uzbekistan', *EIU Country Profile 1995–1996* (London, 1996), p. 118.

[15] Food and Agricultural Policy Unit (FAPU) of the TACIS programme of the European Union, *Marketing and Price Policy*, p. 5.

[16] See Food and Agricultural Policy Unit (FAPU) of the TACIS programme of the European Union, *Land Reform Policy in the Republic of Uzbekistan*, 11 June 1996, Tashkent, for an excellent and detailed presentation and discussion of the major problems.

[17] Ibid., pp. 17–25; Philip P. Micklin, 'Development of Self-governing Irrigation Systems in Uzbekistan: Problems and Prospects', draft final report on the training seminar held in Tashkent, Uzbekistan, April 29–30, 1997, prepared for the Central Asia Mission US Agency for International Development, Almaty, Kazakhstan under Contract No. CCN-0003-Q-14-3165-00 of the Environmental Policy and Technology Project, 25 May 1997.

through the collective's distribution system. The problem is that water deliveries are not guaranteed. When supplies are tight during the most critical plant growth periods of summer, it is routine for the collective farm management to reduce or entirely cut off deliveries. This, of course, lowers yields, and has ruined many private farmers. It is also one of the key factors hindering the development of this sector of the agricultural economy. Cooperative farms in Uzbekistan in league with the local *hakims* have used their formidable power over private farms to force them to act as extensions of the collective – growing cotton and grain to help the collectives meet state orders and the local *hakims* to stay in favour with higher-level government officials.

A third problem for private farmers is the absence of accessible systems for information (for example easy access to topographic, land use and soil maps) and competent advice on proper land and water management.[18] Most private farmers lack knowledge of such subjects. Those who had prior work on the *kolkhoz* were usually not broadly exposed to these matters and the many newcomers to farming have no experience with irrigation at all. The collectives often have technical sections with personnel trained in soils, farming methods and irrigation technology, but private farmers generally do not have access to this expertise.

A final problem for private farmers is the absence of a system of water rights in the former Aral Sea basin republics of the Soviet Union. Farmer-irrigators need to know they have legally assured rights to a certain amount of water on a long-term basis and that these rights will not be abrogated except for well-defined and sound reasons.

6.2.3 Irrigation water pricing

Price is one of the key tools for the allocation of scarce resources, including water, and for ensuring their efficient use. In the 1920s, there was a charge for irrigation water in the Soviet Union, but these payments were dropped with the beginning of collectivization.[19] These tariffs were also instituted from 1949 to 1957. During the last 15 years of the USSR there was renewed interest in irrigation water pricing. Water managers tried several experiments with water pricing in the Aral Sea basin (in Kyrgyzstan and Uzbekistan). On the basis of these, the government intended to introduce universal irrigation water pricing in 1991.

Interest in water pricing has continued in the post-Soviet period. The basin states clearly understand that it could be a very effective means of regulating and improving the efficiency of water use for irrigation and providing funds for the operation and

[18] See Tom Mccray, 'Complicating Agricultural Reforms in Uzbekistan: Observations on the Lower Zaravshan Basin', *Central Asia Monitor*, on-line supplement, No. 2, 1997. (*http://homepages.together.net/~chalidze/cam/sam8.htm*) for a most useful, first-hand account of the problems private farmers face in one of the key irrigation zones.

[19] Philip P. Micklin, *The Water Management Crisis*, pp. 32–6.

upkeep of irrigation systems. Furthermore, the major international donors (USAID, the World Bank, the EU) have made the introduction of such charges a priority in their market reform efforts and have pushed all the basin states to move expeditiously towards this goal.[20]

Kazakhstan and Kyrgyzstan have made the most serious efforts to implement meaningful irrigation water charges.[21] Both states introduced systems of differential payments in 1992. However, the fees (ranging in Kazakhstan from \$0.0067 to 0.067/m^3 and in Kyrgyzstan from \$0.0045 to 0.045/m^3) are not sufficient to cover operation and maintenance (O&M) costs. Furthermore, only a fraction of the money owed by farmers is collected. Consequently, the irrigation systems in both nations continue to deteriorate owing to lack of repair. In August 1997 the government of Kazakhstan adopted a new set of procedures for water charges, including irrigation.[22] However, although an improvement, the new regulations are still unlikely to provide sufficient revenue to cover basic operating costs, let alone generate enough money for investment in rehabilitation and new construction.

The situation in Tajikistan is unclear, although the government's intention was to introduce some sort of irrigation water pricing in May 1996.[23] In Turkmenistan, as a matter of fundamental state policy, irrigation water is provided free to farms. But a charge is assessed for usage above the plan, which the farm is required to submit annually to government water management agencies.[24]

Uzbekistan, the major irrigating nation in the Aral Sea basin, has lagged well behind Kazakhstan and Kyrgyzstan in moving towards irrigation water pricing.[25] The country's legal code has allowed such charges since 1993 for the purpose of limiting water use and providing funds to maintain water delivery systems.[26] Nevertheless, the government provided water without any charge to farms until 1997. Since then, irrigators

[20] Information acquired by the author during a one-year assignment as Resident Advisor on Water and Environmental Management Policy to the Government of Uzbekistan, under USAID's Environmental Policy and Technology Project, September 1996 to October 1997.

[21] R. Burger, *Water Legislation and Pricing in Kazakhstan*, Environment Discussion Paper No. 40, February 1998, NIS-EEP Project, 17 pages; D. Smith, 'New Independent States Report: Comparative Analysis of Water Pricing Policies and Water Allocation Techniques in the Central Asian Republics, 13 March– 23 April 1996, Almaty, Kazakstan, Bishkek, Kyrgyzstan, and Tashkent, Uzbekistan', Delivery Order 08, Task A and C, Environmental Policy and Technology Project, Contract No. CCN-003-Q-00-3165-00.

[22] Burger, *Water Legislation*.

[23] Smith, 'New Independent States', p. 5.

[24] Ibid., p. 6.

[25] Philip P. Micklin,. 'Developing Water Pricing Systems for Uzbekistan: Key Policy Issues and Initial Steps', draft final report on the training seminar held in Khodjikent, Uzbekistan, July 28–1 August 1997, prepared for the Central Asia Mission US Agency for International Development, Almaty, Kazakhstan under Contract No. CCN-0003-Q-14-3165-00 of the Environmental Policy and Technology Project, 11 August 1997.

[26] Ministerstvo yustitsii Respubliki Uzbekistan, *Novyye zakony Uzbekistana* [Ministry of Justice of the Republic of Uzbekistan, New Laws of Uzbekistan], No. 8 (Tashkent: Adolat, 1994), p. 63, article 30.

have been obliged to pay a water tax, which for 1997 was around $0.0001/m^3 for surface water and $0.00013/m^3 for groundwater (based on black market rates of money exchange).[27] This very small charge is not a real incentive for farmers to undertake careful water management. Furthermore, the money collected goes into general revenues and is not specifically designated for O&M of the irrigation systems.

Unquestionably, implementation of meaningful irrigation water pricing schemes in all five of the former USSR basin states, with charges high enough to provide sufficient funds for O&M and an enforceable collection mechanism, would be a major step towards improved water management. On the other hand, there are some fundamental practical barriers to this – government recalcitrance (e.g. in Uzbekistan and Turkmenistan) and the natural resistance from farmers. Many farms, probably the vast majority in the Aral Sea basin, whether cooperatives, leased or peasant, are in such poor financial condition that it would be very difficult for them to pay any substantial fee for water without going bankrupt. Lack of funds also makes it difficult to rebuild dilapidated irrigation facilities to reduce water usage and thus lower payments. Recognition of this is one of the major reasons why the governments have not taken more resolute actions to collect the modest charges now imposed.

Another serious obstacle to the introduction of water pricing is the lack or inadequacy of flow measurements at farm turnouts (points where water is delivered to farms from the main supply canals). Cooperative farms usually do have some method of measuring deliveries to the farm such as a calibrated weir, a staff-gauge from which the height of water can be read or an adjustable gate that can be set to allow a specific amount of water to flow past. But these facilities are often not working or the measurements are not recorded. At the private farm level, measuring devices are almost completely absent and usage is a matter of good guesses, at best.

Commonly, for both cooperative farms and private farms, water usage is calculated by multiplying the area of each crop by the water usage norm for each crop raised (i.e. what should be used, not what is used) and summing the results. Without any means of directly measuring deliveries of water to farms, it is difficult to estimate water usage accurately or to introduce an equitable water charge system.

6.2.4 Irrigation water supply systems

The irrigation water delivery systems in all five of the former Soviet republics still reflect their Soviet origins. They are centralized and hierarchical. At the top are Ministries of Reclamation and Water Management (in Turkmenistan and Tajikistan), Agriculture (in Kazakhstan) and Agriculture and Water Management (in Uzbekistan

[27] 'Pis'mo o poryadke uplaty za vodu v 1997 godu' [Letter on the procedure for water payments in 1997], *Nalozhnyye i tamozhennyye vesti* [Tax and Custom News], No. 12/13 February–19 March 1997, p. 10.

and Kyrgyzstan). Directly subordinate to these organizations are branches at the *oblast'* (administrative region) level. Under them are the *rayon* (administrative district) level offices. These organizations are responsible for and manage all irrigation delivery facilities down to the cooperative farms. As in Soviet times, the systems are run from the top down. Decisions about water allocations and deliveries are made in the central office and then communicated to the *oblast'* branches and from there to the *rayon* organizations which arrange the final supply to the consumers.

These systems were designed for a planned economy where the basic agricultural and water management decisions were in the hands of central planning officials and the national and republican governments provided the funds for building and running the water management system. Since independence and as the new nations of the Aral Sea basin move towards more market-oriented, decentralized governments and economies, these systems are suffering from a number of major difficulties.

Deteriorating infrastructure is a critical issue. Even during Soviet days, poor maintenance of irrigation systems was common.[28] The situation has become far worse in recent years.[29] Facilities such as canals, pumping stations, diversion head-works, drainage facilities and dams are literally falling apart as money for maintenance from governmental budgets shrinks and is not replaced from other sources. Access to critical maintenance equipment and parts, frequently produced in parts of the former Soviet Union outside Central Asia, has also became difficult and expensive.

In Uzbekistan maintenance of inter-farm canals is an especially acute problem. These canals carry water from one cooperative farm to another. Upkeep of the main and distributory irrigation canals is the responsibility of the Ministry of Agriculture and Water Management (*Minsel'vodkhoz*) and its *oblast'* affiliates (*obsel'vodkhoz*), although they lack funding to do a proper job. Repair and servicing of the intra-farm network is the job of the cooperative farm or the private farmer receiving the water (both are also usually too impoverished and technically ill-equipped to do this work adequately). Responsibility for maintenance of the inter-farm network is less clear. It appears to be the job of the *obsel'vodkhoz*, but as these facilities lie off the main canals and between farms they do not receive the attention devoted to the main distribution system. Consequently, these canals are deteriorating even more rapidly than the main and on-farm systems, contributing to excessive flow losses in them and to problems in ensuring farms and farmers at the tail end of the irrigation network receive adequate water supplies.

A final serious problem is that the *rayon*-level offices of the agencies for distributing irrigation water are neither accustomed nor attuned to dealing with private farmers. They were organized to serve collective and state farms. Kyrgyzstan, in what

[28] Micklin, *The Water Management Crisis*, pp. 32–3.
[29] Burger, *Water Legislation*; Micklin, 'Developing Water Pricing Systems'.

may be a model for the other basin states of the former USSR, has made it possible for private irrigators, through agricultural water user associations, to contract directly with these agencies for irrigation water deliveries, ensuring a certain amount of water will be delivered at a specific time for an agreed price.[30] This is in contrast to the situation in Uzbekistan, where private farmers must arrange their water deliveries through the cooperatives with no assurance that these arrangements will be honoured.

6.2.5 Self-governing irrigation systems

A potentially promising solution to the most serious agricultural water management problems facing the new nations of the Aral Sea basin is the formation of self-governing irrigation systems, commonly referred to simply as water user associations (WUAs). These organizations are common in both the developed and the developing world (e.g. India, the United States, Mexico, the Philippines, Egypt and Pakistan) and have proved their worth as means to improve agricultural productivity and water use efficiency by giving farmer-irrigators responsibility for and a clear stake in agricultural and water management at the local level.[31] Information about the utility of these organizations in the Aral Sea basin presented below is mainly based on a seminar on self-governing irrigation systems held in Tashkent, Uzbekistan in April 1997.[32]

In Uzbekistan, Tajikistan and Turkmenistan, where the cooperative farm system remains to one degree or another, such organizations could provide a single point of contact for negotiations, contracts and dispute resolution between private and collective farmers and *oblast'/rayon* branches of the national water management ministries as well as with regional and local authorities. They could also help reduce conflict and promote cooperation between private farmers and the cooperatives by demonstrating the mutual benefits of working together. These organizations could monitor water deliveries within irrigation systems and help prevent farmers at the head end of the system making excessive withdrawals that force those at the tail end to rely heavily on salinized drainage water for their irrigation supply.[33]

In all five new states, WUAs could facilitate the introduction of irrigation water pricing by acting as the purchasing agent buying water directly from the state distribution system and delivering it to private and cooperative farms. A fee would be collected for this service, part of which would go to the state delivery organ and part of which would be kept by the WUA to fund its activities. As noted above, Kyrgyzstan has already successfully experimented with this system.

[30] Visit to Bazarkurgan water users association, Djalabad, Kyrgystan, 9–10 April 1997.

[31] Elinor Ostrom, *Crafting Institutions for Self-Governing Irrigation Systems* (San Francisco: ICS Press, 1992).

[32] Micklin, 'Development of Self-governing Irrigation Systems'.

[33] T. Kamalov, 'Zemlya i voda obreli yedinogo khozyaina' [Land and water have found a single master], *Narodnoye slovo*, 4 April 1997, pp. 1–2.

The associations, acting alone and together, could promote and protect the interests of their members by legal and lobbying measures at the local, regional and national levels. They could develop their own maintenance services for on-farm and inter-farm water distribution networks and raise the money (through member assessments and allocation of part of the water fees collected) to fund these operations. The associations could provide a variety of key technical, supply, marketing and financial (credit) services to private and cooperative farmers. They could also lead the effort for the introduction of improved water use and agricultural technologies to these same groups.

However, for these organizations to flourish requires specific actions from the national governments. The first step would be the implementation of national policies promoting the development of agricultural water user associations through a presidential decree and/or legislation from the parliaments. It is also critical that the rights of these entities be clearly enumerated and legally protected. Finally, to make a system of agricultural water user associations work as it should requires the development of unambiguous systems of water rights. These should be assigned to existing water users on the basis of the beneficial use concept, i.e. that each right holder uses the water assigned to him in an economically beneficial way. Holders of rights should be able to trade and sell these unless a clear and serious harm to other water right holders or the public interest can be demonstrated.

Uzbekistan, Kyrgyzstan and Kazakhstan have legislation and/or decrees allowing the formation of agricultural water user associations. Kyrgyzstan and Kazakhstan take a more liberal, hands-off approach to the operation of these than Uzbekistan. However, none of these states permits these organizations the necessary level of rights and autonomy.

6.3 Corruption and patronage in water management

Corruption and patronage, without question, are common in the new nations of the Aral Sea basin. This is particularly true among the Uzbeks, who have a strong tradition of hierarchy and authoritarianism which survived, some would say flourished, during the Soviet regime.[34] Corruption and patronage have remained strong since independence. However, Western analysts of the Central Asian scene must be cautious in their evaluation of the extent and negative effect of these influences: we often tend to be shocked by these phenomena in other societies, while overlooking the different forms of the same occurrences within our own societies.

[34] G. Gleason, *The Central Asian States: Discovering Independence*, Westview Series on the Post-Soviet Republics (Boulder, CO: Westview, 1997), pp. 117–18; A. Bohr, *Uzbekistan: Politics and Foreign Policy*, Central Asian and Caucasian Prospects Project (London: Royal Institute of International Affairs 1998), pp. 3–9.

Much of our view of corruption and patronage in Central Asia is shaped by the so-called cotton scandals that befell Uzbekistan during the 1980s. After the death of Leonid Brezhnev, first party secretary of the Communist Party of the Soviet Union, in 1982, the new Moscow leadership launched a campaign against crime, corruption and patronage in the republic. The main basis of the campaign was the charge that local Communist Party leaders and functionaries, and government bureaucrats (almost all Uzbeks), had engaged in falsification of cotton production data to show higher production and higher yields than were obtained. They were also accused of large-scale theft of cotton for private sale; embezzlement of state funds; the use of the ill-gotten gains to build lavish dachas, buy expensive cars and otherwise live the good life; the handing out of patronage positions to friends and supporters; and the bribery of police, judges and others 'to look the other way'.[35] The case was touted as '... the biggest in Soviet postwar history in terms of the sums stolen and the political, economic and social damage caused by the crime'.[36]

In a five-year campaign, party and government officials at all levels were removed from power, many of them arrested and imprisoned, and some shot. The anti-corruption effort reached its peak under Gorbachev in the late 1980s. Media reports of the time suggested that corrupt and criminal elements controlled Uzbekistan.

Posthumously, the most famous Uzbek political figure of the Soviet era, Sharaf Rashidov, first party secretary of Uzbekistan and a candidate member of the Politburo in Moscow from 1961 until his death in 1983, was accused of being part and parcel of the corruption in Uzbekistan. Denounced at the 1986 Uzbek party congress, he was stripped of the numerous awards and medals that had been conferred on him, chiefly by his friend Brezhnev.

The campaign against corruption in Uzbekistan ended abruptly in 1989. Islam Karimov became first party secretary and in 1991 was elected president of the new nation of Uzbekistan. He continues in that office today. Karimov inaugurated a new era that stressed Uzbek nationalism.[37] Part of this effort was denunciation of the Moscow-run anti-corruption campaign as politically inspired and very one-sided, along with the rehabilitation of most of those accused of related crimes. Karimov elevated Rashidov to the status of a national hero, extolling him as a brave and wise leader who fought hard for the benefit of Uzbekistan.[38]

[35] 'Investigations Continue', *Moscow News*, No. 44, 1988, p. 14; 'Up against the Mafia', *Moscow News*, No. 14, 1988, p. 13; 'Requital', *Moscow News*, No. 51, 1988, p. 13; 'Kolonna', *Pravda Vostoka*, 19 July 1988, p. 4.

[36] 'Up against the Mafia', p. 13.

[37] N. Mishin, *Islam Karimov –The First President of the Republic of Uzbekistan* (Tashkent: Uzbekiston, 1997), pp. 66–8.

[38] 'K 75 letiyu so dnya rozhdeniya Sh. R. Rashidova' [On the 75th anniversary of the birth of S.R. Rashidov], *Pravda Vostoka*, 26 September 1992, p. 1.

Post-independence economic and political conditions in the former USSR Aral Sea basin states are, without doubt, conducive to corruption and patronage in all spheres of activity, including water management. Officials and the general populace commonly expect supplementary payments for services that are outside any formal agreement or contract. The centralized systems of water control and distribution, little changed since Soviet times, make it easy for those in charge to ask for and receive something extra for delivering irrigation water on time and in the necessary quantity.

Regional and local political leaders and administrators, particularly in Uzbekistan and Turkmenistan, are powerful. As discussed in Chapter 5, they have considerable sway over both cooperative farmers and private farmers in terms of land use and water distribution. Hence, it is easy for the administrators to extract payments from farmers for favourable treatment or for refraining from activities harmful to farmers. Fear of corruption has been one of the major reasons for periodic replacement of regional and local *hakims* in Uzbekistan.[39]

The governments of the Central Asian states have to some extent struggled to control corruption. Nevertheless, some of the same corruption problems that existed during the 1980s have again appeared in Uzbekistan: bribe taking, abuse of office, the illegal sale of cotton to foreign countries and the illegal export of defoliants.[40] In some cases, authorities have taken criminal action against the perpetrators, and the Minister of Agriculture and Water Management was given early retirement in 1998.

Patronage is common throughout Central Asia but particularly in Uzbekistan with its powerful clans. Political leaders at all levels hand out positions and opportunities to their own people not on the basis of competence, training, experience or ability but according to personal allegiance.

The influx of the donor community to the Aral Sea basin is also a factor promoting corruption and patronage. The big donors (the World Bank, USAID and the EU) have injected large amounts of hard currency into various Aral Sea water management improvement efforts. Much of this money is funnelled through the individual governments or through the regional organizations that have been created to handle Aral Sea basin issues. Although it is difficult to know the scale, some of this money is being siphoned off as it passes through the bureaucratic channels of regional and governmental agencies.[41] The fact that Western consultants working in these programmes earn salaries that are on average probably tenfold those of their local counterparts, often for the same work, legitimizes for locals the idea that it is all right to skim off some extra money for themselves. Western companies bidding on contracts through the Aral Sea basin improvement programmes have been asked to pay bribes to key officials to improve their chances of winning.

[39] Bohr, *Uzbekistan: Politics and Foreign Policy*, 1998, p. 5.
[40] Ibid., pp. 23–4.
[41] Information acquired by the author during a one-year assignment.

Corruption and patronage hinder reform in water management as they do in other economic sectors. Money that should be used for improving irrigated agriculture and raising the efficiency of water use goes into someone's pocket. Farmers are forced to pay local officials for services they should receive as a right. People are hired not on the basis of merit but on the basis of their connections. Furthermore, corruption causes disillusionment among the many honest, hardworking people in these nations who want a transparent and honestly run system, and it creates a sense that efforts to institute real change are useless. However, corruption and patronage are at heart symptoms of the underlying barriers to reform: over-centralized, authoritarian governments; tightly controlled economies; lack of democracy; entrenched, anti-reform bureaucracies; and mistrust (on the part of governments) of people's ability to run their own affairs properly.

7 THE FUTURE OF WATER MANAGEMENT

The Aral Sea basin is not inherently short of water resources. The flow of the Amu Dar'ya and Syr Dar'ya on an average annual basis and renewable supplies of groundwater are substantial and sufficient to meet current withdrawals. On the other hand, average figures are deceptive and do not reflect the intra- and inter-year variability of these rivers' discharge and the associated problem of storing flow during surplus flow seasons and years for use during periods of flow deficit, when demand is greater than natural availability. When these factors are taken into consideration, and given the current usage level, water resources are strained, particularly in the cycles of low water years.

The Aral Sea basin states face a plethora of problems in the management of their water resources. The key problem is irrigated agriculture. This activity is the economic anchor for the countries of the region and accounts for more than 90 per cent of water withdrawals in the five new states. Uzbekistan and Turkmenistan are the chief irrigating nations in the basin. All states apart from Kazakhstan intend to expand the irrigated area. For one state to increase irrigation withdrawals from the Amu Dar'ya and/or Syr Dar'ya requires the other states along these rivers to reduce their usage.

One alternative would be the implementation of a large-scale rehabilitation and modernization programme for irrigation systems that could improve efficiency of water use and free water for irrigation expansion. A comprehensive, thorough programme to raise water use efficiency in irrigation could perhaps release 28 per cent of the 100 km^3 or so withdrawn for this activity in 1995. This would require massive investment, at least $16 billion, which is highly unlikely to be made in the near- to medium-term future. Crop substitution is probably a much cheaper means to obtain sizeable reductions in water use, but if undertaken on a large scale it would necessitate a considerable reduction in the hectarage devoted to cotton, the chief export crop for Uzbekistan and Turkmenistan and a foreign currency earner for the former.

Population growth and global climate change are two other factors which work against stabilizing or reducing water use in irrigation. The population of the basin is projected to grow to 60 million in 2020, a 28 per cent increase on the 1996 figure of 47 million.[1] The need to provide employment and food for a much larger population

[1] Tashkent Institute of Engineers of Irrigation and Agricultural Mechanization and the Aral Sea International Committee, 'The Mirzaev Report', May 1998, Table 1.

could require (and the basin governments say will require) a significant increase in the irrigated area and larger water withdrawals. The majority of experts believe global climate change towards perceptibly warmer conditions from the anthropogenically caused release of greenhouse gases is a near certainty in the new century. Climate models are not yet sufficiently sensitive to indicate with reliability the regional consequences for areas the size of the Aral Sea basin. But the result could well be significant warming that will increase crop water needs and act as another force for increased water usage in irrigation.

The adoption of fundamental reforms is crucial to improving water management. The major international donors such as the World Bank, the EU and USAID have been unanimous in urging the Aral Sea basin states to proceed along this path. The introduction of irrigation water pricing that is meaningfully related to the costs of providing the resource, the true privatization of land and the granting of rights of self-governance and responsibility for management of irrigation systems to farmer-irrigators are the most needed steps to encourage farmers to use water more carefully. A partial return to the use of small-scale irrigation systems that characterized parts of Central Asia in earlier times, coupled with the sensible employment of modern irrigation technology (e.g. laser levelling of fields) could also contribute to water savings.

There is no denying the progress that has been made in the reform effort. Kazakhstan and Kyrgyzstan have abolished state and collective farms and are proceeding towards true privatization of land and the placing of agricultural decision-making in the hands of farmers. These states have also introduced irrigation water pricing. Uzbekistan has ended the state farms system, given somewhat greater freedom of action to the cooperatives, allowed the development of certain types of (partially) private farming and abolished state orders for all crops, including cotton. Tajikistan has stated an intention to convert state and collective farms to cooperatives and to encourage private-sector agriculture. Kazakhstan, Kyrgyzstan and Uzbekistan are encouraging the formation of agricultural water user associations to accept responsibility for the management and allocation of irrigation water at the local level.

Nevertheless, serious problems remain. Farmers are impoverished throughout the basin and most cannot afford to pay more than a token charge for water. Kazakhstan and Kyrgyzstan are facing significant opposition from rural constituencies to further land privatization and the reduction of state control of agriculture. Land in Uzbekistan and Turkmenistan remains state property, and there are only vague plans for true privatization. Both governments still exert a strong hand over agriculture and water management, reminiscent in many ways of the former Soviet system. Uzbekistan's leadership has articulated some legitimate reasons for going slow on reform – concerns related to protecting the interests and welfare of the rural population in densely settled regions; avoiding excessive land speculation and land grabbing by the rich; trying to avoid ethnically based conflicts over land and water as have occurred in the Fergana valley; preserving prime agricultural land from conversion to non-agricultural uses; and protecting the monopsony on cotton – the government's main source of foreign exchange.

Corruption and patronage remain endemic, society-wide problems, particularly in Uzbekistan. These are a hindrance to reform efforts in the agriculture and water management sectors as they divert funds from their most productive uses, discourage reform efforts at the local level and block competent people from advancing in the institutional structure of the agriculture and water management sectors.

Another serious area of concern is the sharing of the international rivers of the Aral Sea basin. Inherent water management conflicts exist between the upstream states (Tajikistan, Kyrgyzstan and, to a lesser extent, Afghanistan) and the downstream states (Uzbekistan, Kazakhstan and Turkmenistan). The former generate most of the basin river flow but use relatively little, while the situation is reversed for the latter. Tajikistan, and particularly Kyrgyzstan, are intent on withdrawing significantly more of the water generated from their territory to expand irrigation. This action is opposed by Uzbekistan, Kazakhstan and Turkmenistan as it reduces the water available for their national uses. The upstream states also want to operate the large hydroelectric stations located on their territory to maximize winter hydropower production, which is counter to the interests of downstream irrigating states that need maximum releases during the summer. Sharp conflicts over this issue have arisen between Uzbekistan and Kazakhstan on the one hand and Kyrgyzstan on the other. Serious differences have also developed among the downstreamers. Relations are particularly strained between Uzbekistan and Turkmenistan over the latter's plans to increase diversions to the Kara-Kum canal.

The situation is exacerbated by two other factors. First, since independence, river flow in the basin has been well above the long-term average, giving the new nations a false sense of security as to water availability (see footnote 28, Chapter 3). The inevitable shift back towards average or below average hydrologic conditions, and less water for sharing, could heighten interstate tensions. Making matters even worse is the basin states' tendency to overestimate their rightful share of water resources. These exceed any reasonable estimate of the usable basin-wide resource.[2]

Finally there is the problem of the Aral Sea. It continues to shrink and the serious environmental and human problems in the surrounding disaster zone are not visibly abating. To provide sufficient inflow to stabilize the large sea in the south, let alone begin to raise its level, would require sizeable cutbacks in irrigation (or a major and costly programme to improve irrigation efficiency and implement crop substitution). None of these options is likely to be implemented in the foreseeable future. The former Soviet basin states have formally agreed to provide more water in order to

[2] World Bank et al., p. 15; ICAS, *Fundamental Provisions*, Chapter 6.

preserve what is left of the deltas of the two rivers and partially restore the now separated northern Aral Sea, but attaining even these modest goals is proving difficult.

In spite of problems, improved management of interstate waters in the Aral Sea basin is far from being a lost cause. Since independence, the basin states have established several basin-wide organizations to facilitate cooperation in coping with the problems of water sharing and the Aral Sea. The international community, through the main multilateral and bilateral aid organizations, is providing financial and technical assistance plus policy advice to help in these endeavours. This represents a major contribution to conflict resolution in the region. However, progress is slow and uneven, and the most critical problems remain largely unresolved.

Looking to the future, the only rational avenue for the Aral Sea basin states to follow is cooperation and compromise in managing and sharing their transnational water resources. This is necessary not only to avoid interstate conflict, perhaps even military confrontation, but to develop an integrated, basin-wide strategy to optimize water use efficiency and maximize efforts to restore and protect key water-related ecosystems such as the Aral Sea and its environs. Such a strategy would benefit all basin riparians. Indeed, the abundantly obvious need for cooperation on interstate water management issues may play a key role in avoiding conflict and promoting interstate cooperation on other matters. The attitude of Uzbekistan, as the basin state that is the most populous, politically powerful, economically developed and strategically situated, as well as the heaviest user of irrigation, is crucial to successful interstate cooperative management of transnational waters. If this state focuses excessively on national self-interest in water management, as it has been wont to do in the past, regional efforts for the effective and peaceful management of both national and transnational water resources will founder.

FURTHER READING

Books and articles

Craumer, Peter, *Rural and Agricultural Development in Uzbekistan*, Former Soviet South project key paper (London: Royal Institute of International Affairs, 1995).

Glantz, Michael H. (ed.), *Creeping Environmental Problems and Sustainable Development in the Aral Sea Basin* (Cambridge: Cambridge University Press, 1999).

Gleason, Gregory, *The Central Asian States: Discovering Independence*, Westview Series on the Post-Soviet Republics (Boulder, CO: Westview, 1997).

Lewis, Robert A. (ed.), *Geographic Perspectives on Soviet Central Asia* (New York: Routledge, 1992).

Micklin, Philip P., *The Water Management Crisis in Soviet Central Asia*, The Carl Beck Papers in Russian and East European Studies, No. 905 (Pittsburgh, PA: The Center for Russian and East European Studies, August 1991).

Micklin, Philip P. (ed.), *Special Issue on the Aral Sea Crisis, Post-Soviet Geography* (formerly *Soviet Geography*), Vol. XXXIII, No. 5 (May 1992).

Micklin, Philip P., 'Regional and International Responses to the Aral Crisis,' *Post-Soviet Geography and Economics*, Vol. 39, No. 7 (September 1998), pp. 399–417.

Micklin, Philip P. and Williams, William D. (eds), *The Aral Sea Basin*, Proceedings of an Advanced Research Workshop, May 2–5 1994, Tashkent, Uzbekistan, NATO ASI series, Vol. 12 (Heidelberg: Springer Verlag, 1996).

Smith, David R., 'Environmental Security and Shared Water Resources in Post-Soviet Central Asia, *Post-Soviet Geography* (formerly *Soviet Geography*), Vol. 36, No. 6 (June 1995), pp. 351–70.

Thurman, Michael J., 'The "Command-Administrative System"', *in Cotton Farming in Uzbekistan 1920s to the Present*, Papers on Inner Asia, No. 32 (Bloomington, IN: Indiana University Research Institute for Inner Asian Studies, 1999).

Web sites

ICWC (Interstate Commission on Water Coordination) Central Asia: *www.sicicwc.8m.com*

The Aral Sea Home Page: *www.dfd.dlr.de/app/land/aralsee*

Doctors without Borders – Aral Sea Programme: *www.msf.org/aralsea/index.htm*